Edward Nettleship

The Student's Guide to Diseases of the Eye

Edward Nettleship

The Student's Guide to Diseases of the Eye

ISBN/EAN: 9783744718608

Printed in Europe, USA, Canada, Australia, Japan

Cover: Foto ©berggeist007 / pixelio.de

More available books at **www.hansebooks.com**

THE

STUDENT'S G[UIDE]

TO

DISEASES OF T[HE EYE]

BY

EDWARD NETTLESH[IP]

OPHTHALMIC SURGEON TO ST THO[MAS]

LONDON

J. & A. CHURCHILL, NEW BURLINGTON STREET

1879

To

JONATHAN HUTCHINSON,

Consulting Surgeon to the Moorfields Ophthalmic
Hospital, Senior Surgeon to the London
Hospital, &c.,

THIS

BOOK IS DEDICATED

IN GRATEFUL ADMIRATION OF HIS EMINENT QUALITIES

AS A

Clinical Teacher and Investigator

PREFACE

THE aim of this little book is to supply students with the information they most need on diseases of the eye during their hospital course. It was apparent from the beginning that the task would be a difficult one, all the more as several excellent manuals, covering nearly the same ground, are already before the public. That no one of them singly appeared exactly to cover the ground most important for the first beginner in clinical ophthalmology encouraged me to attempt the present work.

The scope of the work has precluded frequent reference to authors, those named being chiefly such as have made recent additions to our knowledge in this country. I am greatly indebted to Dr Gowers, Dr Barlow, and other friends for much information, and for many valuable suggestions. My best thanks are due to Mr A. D. Davidson for his kind assistance in reading the sheets for the press.

4, WIMPOLE STREET;
October, 1879.

The following are the standard works of which most use has been made; but many original papers and smaller treatises have also been consulted:

Soelberg Wells, 'Treatise on Diseases of the Eye.'

Stellwag, 'Treatise on Diseases of the Eye.'

Graefe and Saemisch, ' Handbuch der Augenheilkunde,' particularly the volumes on Diseases of the Conjunctiva (Saemisch); Diseases of the Retina and Optic Nerve (Leber); Glaucoma (Schmidt); and the Relation of General Diseases to Diseases of the Eye (Förster).

Wecker and Landolt, 'Traité complet d'Ophthalmologie,' vol. i, 1879.

Wecker, 'Chirurgie Oculaire,' 1879.

Donders, ' Anomalies of Accommodation and Refraction of the Eye' (New Sydenham Society's Translation).

Liebreich, ' Atlas d'Ophthalmoscopie.'

Wecker and Jaeger, ' Atlas d'Ophthalmoscopie.'

Magnus, ' Atlas Ophthalmoscopischer.'

Pagenstecher and Genth, ' Atlas of the Pathological Anatomy of the Eyeball.'

Poncet, 'Anatomie Pathologique de l'Oeil,' in Perrin and Poncet's Atlas, 1879.

Schreiber, 'Veränderungen des Augenhintergrundes bei internen Erkrankungen,' 1878.

Gowers, 'Manual and Atlas of Medical Ophthalmoscopy,' 1879.

Clifford Allbutt, 'The Ophthalmoscope in Diseases of the Nervous System and Kidneys.'

CONTENTS

PART II.—CLINICAL DIVISION.

CHAPTER IV

DISEASES OF THE EYELIDS

CHAPTER V

DISEASES OF THE LACHRYMAL APPARATUS

CHAPTER VI

DISEASES OF THE CONJUNCTIVA

CHAPTER VII

DISEASES OF THE CORNEA

CHAPTER XIV

DISEASES OF THE RETINA   PAGE

CHAPTER XV

DISEASES OF THE VITREOUS

CHAPTER XVI

GLAUCOMA

CHAPTER XVII

DISEASES OF THE OPTIC NERVE

CHAPTER XX

STRABISMUS AND PARALYSIS

CHAPTER XXI

OPERATIONS

PART III.—DISEASES OF THE EYE IN RELATION
TO GENERAL DISEASES

CHAPTER XXII

A. General Diseases

Eye diseases caused by :—Syphilis, acquired and inherited,
and diseases of optic nerve in relation to syphilis; Smallpox ; Scarlet fever, typhus, &c ; Diphtheria ; Measles ;
Chicken-pox and whooping-cough ; Malarial fevers ;
Relapsing fever ; Epidemic, cerebro-spinal Meningitis ; Purpura and scurvy ; Lead poisoning ; Alcohol ;
Tobacco ; Quinine ; Kidney disease ; Diabetes ; Leucocythæmia ; Pernicious anæmia ; Heart disease ;
Tuberculosis ; Rheumatism and gonorrhœal rheumatism ; Gout, personal and inherited ; Struma ; Entozoa.

B. Local Disease at a Distance from the Eye

Eye symptoms caused by :—Megrim ; Neuralgia ; Hysteria.
Disease of brain : Cerebral tumour ; Syphilitic disease ;
Meningitis ; Cerebritis ; Hydrocephalus ; Convulsions.
Disease of spinal cord ; General paralysis of insane.
Motor disorders of eyes in cerebral and spinal dis-
ease.

C. The Eye sharing in a Local Disease of the Neighbouring Parts

Eye symptoms caused by :—Herpes zoster of fifth nerve ;
Paralysis of fifth nerve, of facial, and of sympathetic ;
Exophthalmic goitre ; Erysipelas and orbital cellulitis.

PART I

MEANS OF DIAGNOSIS

CHAPTER I

LEADING SYMPTOMS

A PATIENT who seeks advice about his eyes will generally lay stress upon (A) some external inflammatory or irritative symptoms; (B) defective sight alone; (C) a squint or some other disfigurement, or inability to open or shut the lids; (D) pain may be the chief or only symptom.

(A.) There are symptoms of *external irritation or inflammation.*—(1.) There is *watering, photophobia, or swelling of the lids,* with or without redness of the eye, and defective sight. Examine the lids for ophthalmia tarsi, styes, irregular growth of the lashes, meibomian and other tumours. Look at the inner surface of the lower lids; if there be thickening and redness, evert and examine also the upper lid (for granular disease). Examine the surface of the eyeball carefully as to degree and character of congestion, whether local or general, affecting the exposed or only the covered parts, due to superficial, tortuous, bright red vessels (conjunctivitis), or to deeper, straighter, pink or purplish vessels (ulcers of cornea, iritis, cyclitis). Spots or pustules with local congestion are present (phlyctenular ophthalmia). The cornea shows one or more spots or patches of opacity (*see*

1

Abrasions, foreign bodies and ulcers of cornea), or is hazy all over (keratitis), or shows a number of minute dots at its lower part (keratitis punctata). The chief symptom is persistent watering of one eye —lachrymal obstruction, &c., slight soreness, blinking, a little watery or gummy discharge, and inability to use the eyes for long or to bear bright light or strong wind (hypermetropia, myopia, asthenopia).

(2.) The chief symptom is *discharge with congestion* of the eye and lids (*see* Ophthalmia); or spasmodic closure of the lids and photophobia (*see* Corneal Diseases).

(3.) *Defect of sight* is especially complained of, with more or less inflammatory signs and with or without severe pain. Examine the cornea for ulcers and other haziness; the pupil for size, mobility, and clearness; and the iris for colour and general appearance (iritis, glaucoma). Feel the tension and ascertain roughly the condition of the visual field and note accurately the near and distant sight. Enquire as to injury.

(B.) The complaint is of *defective sight in one or both eyes without other symptoms.* Such symptoms, whether symmetrical or unilateral, may indicate an error of refraction or accommodation, or opacity of some of the media, or diseases of the choroid, retina, or optic nerve.

Ascertain whether one or both eyes are affected; the duration of the defect; and under what circumstances as regards the distance of objects and brightness of light it is most observed. Remember that defect of one eye often remains undiscovered for years until attention is accidentally drawn to it. Is the failure of sight related to bad health or to pain in the head? (albuminuric retinitis, optic neuritis, or atrophy, megrim). Take the near and distant vision, examine the transparency of the cornea, lens (*see* cataract), and vitreous, the colour and appearance of the irides and the size, shape, and mobility of the pupils, and try the tension of the eyes. If the media are clear, the iris and pupil healthy, and the ophthalmoscopic appearances natural, examine the refraction and accommodation (hypermetropia, myopia, presbyopia, paralysis of accommodation). Remember that presbyopia with hypermetropia causes great defect both for distant and near objects. If opacity of vitreous or lens be suspected or proved, or if the defect of sight be not remedied by glasses, it is usually best at once to dilate the pupil with atropine and make a more thorough ophthalmoscopic examination. It is necessary to examine the fundus carefully by the erect image in all doubtful cases. If the disc appears markedly oval either before or after the use of atropine, astigmatism is suspected. As to atropine, it is, as a rule, far better to use it (to the extent of dilating the pupil) than by examining with a small pupil to run the risk of overlooking small but important changes in the lens, vitreous, or fundus. The necessity for its use will depend very much on the skill of the observer and on how much time he can spend over the case; for the larger the pupil the more easily and quickly is the fundus explored. When the sight is pretty good the patient should always be warned that the atropine will dilate the pupil and make the sight dim for a day or two. When there are changes in the optic disc or reason to suspect disease of the optic nerve, the colour perception should be tested. If the complaint is of double vision, ascertain whether it is binocular or monocular by closing one eye; monocular diplopia or polyopia is rare, and is recognised by the persistence of the symptom when one eye is closed.

(c.) There is a squint or some other disfigure-

ment or inability to open the lids (ptosis), or to close
them (paralysis of facial nerve); or defective movement of the globe in one or other direction; or prominence of one or both eyes (proptosis, Basedow's
disease); or the eyelids are swollen but not inflamed
(emphysema, orbital tumours). In myopia the eyes
are often prominent, and if the myopia is one-sided,
this appearance may be unsymmetrical.

(D.) *Pain* is the chief or only symptom complained
of. Note whether it is referred to the eyeball or to
the forehead or temple, &c.; whether periodic and
unrelated to use of the eyes (neuralgia), or irregular
in onset and related to health (megrim), or distinctly
related to use of the eyes (see myopia, hypermetropia, asthenopia). Is there tenderness on pressure
over the supraorbital notch? Test the sight carefully and make a careful ophthalmoscopic examination in all cases; atropine, as a rule, will not be
needed.

The following abbreviations will be used:

T. Tension of the eyeball.
M. Myopia.
H. Hypermetropia.
H. m.   Manifest hyperme-
    tropia.
H. l. Latent hypermetropia.
Pr. Presbyopia.
As. Astigmatism.
A. Accommodation.
V. Acuteness of vision.
p. Punctum proximum or near
    point.

r. Punctum remotum or far
    point.
'. Sign for a foot.
". Sign for an inch.
D. Dioptric, the unit in the
    metrical system of mea-
    suring lenses. A dioptric
    (1 D.) is a lens whose
    focal length is one
    mètre.
y. s. Yellow spot of the re-
    tina.

CHAPTER II

EXTERNAL EXAMINATION OF THE EYE

(1.) To detect *irregularity of the corneal surface,*
make the patient follow some object, *e. g.* the uplifted
finger, moved slowly in different directions, and watch
the reflection of the window from the cornea ; it will
be suddenly broken by any irregularity, such as a
small abrasion.

(2.) To estimate the *tension of the eyeball* (T.) the
patient looks steadily down, and gently closes the eyelids. The observer makes light alternate pressure
on the globe through the upper lid with one finger of
each hand, as in trying for fluctuation, but much
more delicately, and placing the fingers much nearer
together. The finger tips are placed very near
together, and as far back over the sclerotic as possible. The pressure must be gentle, and be directed
vertically *downwards, not backwards.* It is best for
each observer to keep to one pair of fingers, not to use
the index at one time and the middle finger at
another. Patient and observer should always be in
the same relative position, and the best is for both
to stand and face one another. Always compare
the tension of the two eyes. Be sure that the second
eye does not move (roll upwards) during examination, for if this occur a wrong estimate of the tension
may be formed. Some test both eyes at once with
two fingers of each hand. Normal tension is expressed by T. n. ; the degrees of increase and decrease
being indicated by the + or — sign, followed by
the figure 1, 2, or 3. Thus T. + 1 means decided
increase ; T. + 2, much increased, but sclerotic can
still be indented ; T. + 3, eye very hard, cannot be

indented by moderate pressure; T. — 1 — 2 — 3
indicate successive degrees of lowered tension. A
note of interrogation (T. ? + or ? —) for doubtful
cases, and T. n. for the normal, give nine degrees,
which may be usefully distinguished. Equally good
observers often differ in regard to the minor changes
of tension. Apart from variations in delicacy of
touch it is to be remembered that eyes deeply set
in the orbits are more difficult to test, and that T.
in a few cases really does change at short intervals,
e. g. within half an hour. Thickening and stiffness of
the sclerotic alters the apparent tension; the eye
may really be less distended than in health (*i. e.*
T. be —), and yet the want of flexibility of the
sclerotic may give an opposite impression. When
an old blind eye contains bone it feels like wood
covered with washleather.

(3.) The *mobility of the eyeball* may be impaired
in any direction, all directions, and in any degree up
to absolute fixity. Commonly only one eye is affected.
First direct the patient with both eyes open to look
strongly, or follow some upheld object, moved in each
of the four cardinal directions (up, down, right, left);
and next to look at an object (finger or pencil) in the
middle line and rather below the horizontal, which is
gradually approached from 2' to about 6" to test the
convergence power. Of course, in each position we
must notice both eyes; thus, when the patient looks to
his right we have to note the outward movement of his
right and the inward movement of his left. The fixed
marks for inward and outward movements are the inner and outer canthi, and as the apparent freedom of
movement judged in this way varies a little in different
people, symmetrical movements of the two eyes of
the same patient should always be compared. In
looking strongly outwards, the corneal margin usually
reaches, or very nearly reaches, the outer canthus, and
the inner margin the inner canthus in inward movement. In children and stupid people the movements

are often defective unless some patience is used in the examination. In very myopic patients the eyes move somewhat less freely in all directions. The upward movements are estimated by observing the position of the cornea in relation to the border of the lower lid. The upper lid is a less trustworthy guide, since there may be some ptosis or other cause of inequality between the two sides.

(4.) *Squint or strabismus* exists if the visual axes are not both directed to the object of attention, *i.e.* if the rays falling on the yellow spot of one eye do not proceed from the same object as those which fall on the yellow spot of the other. A squint may be the result either of overaction, or of weakness or paralysis of a muscle: the internal rectus by excessive contraction often causes convergent squint ; all other forms, whether convergent or not, result from actual defect of nervous or muscular power.

When a squint is well marked there is no difficulty in identifying the squinting eye as the one which is not directed towards any object held up to the patient's attention. In most cases the patient always squints with the same eye, but in a few he will use either indifferently (alternating squint). Nor is there often. any doubt as to whether the squint is internal (convergent) or external (divergent), *i.e.*, whether the axis of the squinting eye crosses that of its fellow between the patiént and the object of his regard, or crosses it beyond this object or even positively diverges from it. Upward or downward squint, though less common, is quite as evident. To prove beyond doubt which is the squinting eye direct the patient to look at a pencil held up in the middle line at about 18″ from his face, and with a card or piece of ground glass cover the apparently sound, or working eye ; the squinting eye will at once move and look at the pencil, proving that it had previously been misdirected. If the sound eye be watched behind the screen it will be seen to squint as soon as the defective eye "fixes" the object; this is known as the "secondary" squint, and its direction is the same as that of the "primary" or original squint. Thus, if the "primary" squint is convergent, the secondary will also be convergent. In squint from overaction or from mere disuse of one muscle, the secondary and primary deviations are equal. But in a paralytic case (e.g., paralysis of the left external rectus, causing convergent squint of the left eye), the secondary much exceeds the primary squint; for, in order (when the right eye is covered) to move the left eye outwards into the position necessary for fixing the object, a greater *effort* is required than if the sixth nerve were sound; and the same effort being unconsciously transmitted to the healthy right internal rectus, results in a larger movement. In every case of squint it is necessary to test the mobility of the eyes, and to note whether the squint is constant or only occasional.

(5.) *Diplopia* (*double sight*) is almost always caused by a squint, but in many of the most troublesome cases this is so slight as to cause no perceptible deviation of the faulty eye. Diplopia is almost always binocular, disappearing when one eye is covered; but it is occasionally monocular, the patient seeing double with one eye. The latter symptom should be carefully tested before acceptance, and when verified it should lead to examination of the lens and retina, and for nervous symptoms.

To find out what defect of movement is causing binocular diplopia, take the patient into a dark room, and standing at a distance of about 6—8', ask him to follow with his eyes the motion of a candle placed successively in different positions, and to describe the relative places of the double images in each position. Ascertain which of the two images belongs to each eye by placing before one eye a strongly coloured glass and asking which flame is coloured, or by covering one eye and asking which image disappears. In many cases the image formed in the deviating eye (the "false" image) is less bright or distinct, and this difference gives a valuable means of distinguishing the sound from the affected eye; but the patient does not always notice any such difference between the two images, and it is then not always easy to be sure which eye is at fault. The patient's replies should be recorded on a diagram such as is shown in the chapter on Strabismus; the radii may of course be increased for intermediate positions. The false image is marked by the dotted line, the true one by the full line. We have thus a graphic representation of the candle as it appears to the patient, and can deduce from the apparent position of the false image what movements of the corresponding eye are at fault, and, consequently, which muscle or muscles are defective. It is *essential that the patient should not move his head* during the examination, and that he be throughout at the same distance from the candle. Remember that, in the extreme lateral movements, the nose interferes and eclipses one image. When the double images are very wide apart the patient sometimes fails to notice the false image.

For practical purposes it is often enough in a case of diplopia to ask in which directions the double sight is most troublesome, and how the images appear in respect to height, lateral separation, and apparent distance from the patient.

(6.) *Protrusion of the eye (proptosis) and enlargement.*—When unequal the relative prominence of the two eyes is best ascertained by seating the patient in a chair, standing behind him, and comparing the summits of the two corneæ with each other, and with the bridge of the nose. The appearance of prominence or recession, as seen from the front, depends very much on the quantity of sclerotic exposed; thus, slight ptosis gives a sunken appearance to the eyes, and in slight cases of Graves' disease the proptosis seems to increase when the upper lids are spasmodically raised. It is to be remembered that real prominence of the eye may depend on enlargement of the eyeball (myopia, staphyloma, intraocular tumour), as well as on its protrusion, and that if only one eye is myopic, the appearance will be unsymmetrical. In hypermetropia where the eye is too short, and in paralysis of the cervical sympathetic, the eye often looks sunken. Lastly, decided proptosis may follow tenotomy or paralysis of one or more orbital muscles.

(7.) Information derived from *the blood-vessels visible on the surface of the eyeball.*—Three systems of vessels have to be considered in disease, all, owing to their small size, being very imperfectly recognisable in health (Figs. 1 and 2). (1) The *vessels proper to the conjunctiva (posterior conjunctival vessels),* in which it is not important to distinguish between arteries and veins. (2) The *anterior ciliary vessels,* lying in the subconjunctival tissue, and which, by their perforating branches, supply the sclerotic, iris, and ciliary body, and receive blood from Schlemm's canal and the ciliary body: the perforating branches of the *arteries* are seen in health as several rather large tortuous vessels, which stop short about $\frac{1}{12}''$ or $\frac{1}{8}''$ from the corneal margin; their episcleral non-perforating branches are very small and numerous, invisible in health, but when distended forming a pink zone of fine, nearly straight, very closely-set vessels round the cornea ("ciliary congestion," "circum-corneal zone," see Iritis and Diseases of Cornea); the perforating *veins* are very small, but more numerous than the arteries, and their episcleral twigs form a closely-meshed network. (3) The vessels proper to the margin of the cornea, and immediately adjacent zone of conjunctiva (*anterior conjunctival vessels* and their *loop-plexus on the corneal border,* Fig. 1, *l*); by these numerous vessels, which are really offshoots of the anterior ciliary, the two former systems anastomose.

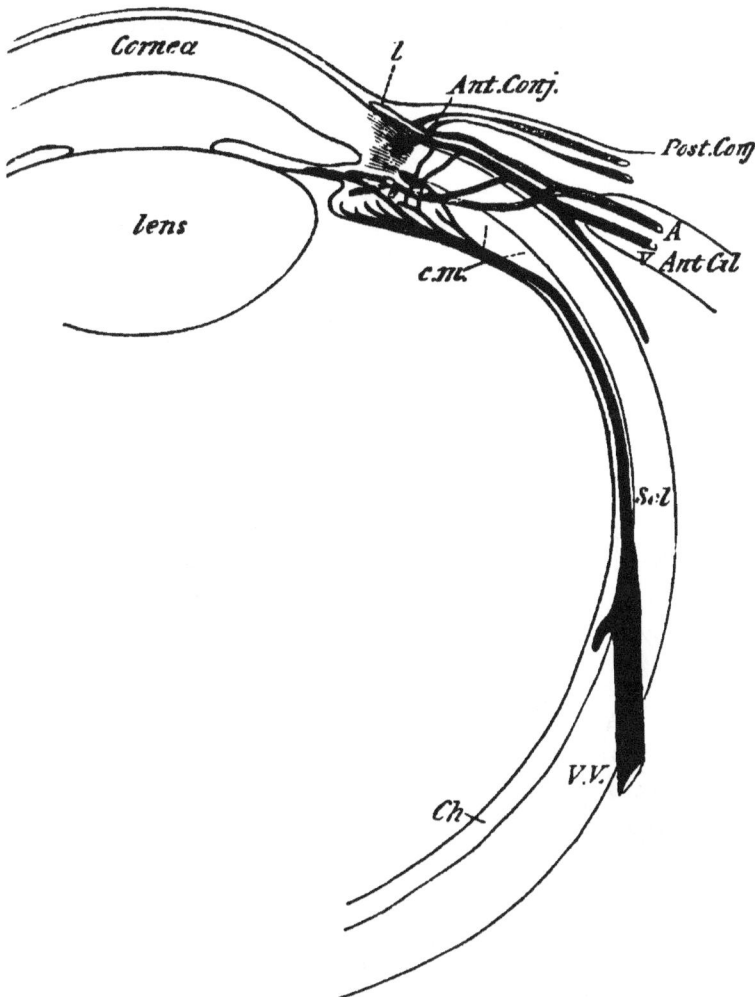

FIG. 1.—Diagram of the vessels of the front of the eyeball.
c. m. Ciliary muscle. *Ch.* Choroid. *Scl.* Sclerotic. *V.V.*
Vena vorticosa. *l.* Marginal loop-plexus of cornea. *Ant.*
and *Post. Conj.*, anterior and posterior conjunctival vessels.
Ant. Cil. A. and *V.*, anterior ciliary arteries and veins;
their episcleral branches are represented too large. (Sim-
plified and altered from Leber.)

Speaking generally, congestion composed of
tortuous bright (brick-red) vessels (System 1),
which move with the conjunctiva when it is slid
over the globe, and which is least intense just

around the cornea, indicates a pure conjunctivitis

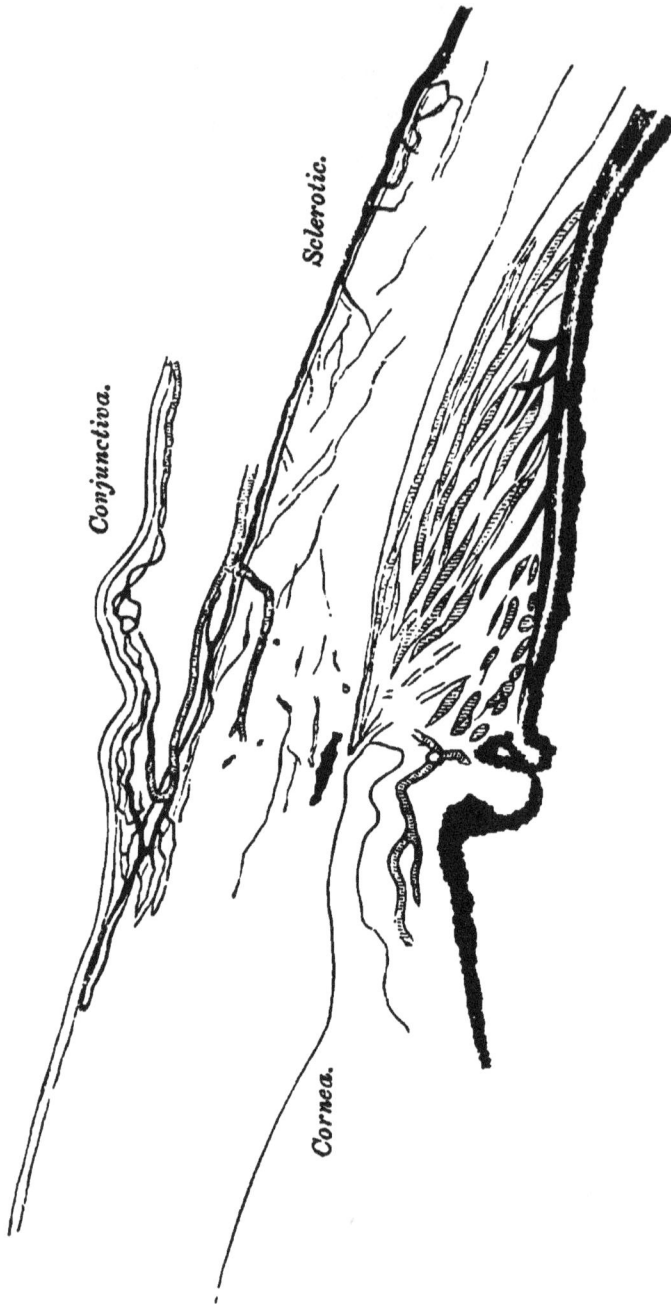

FIG. 2 (more highly magnified) is from a section of this region in an injected human eye showing the actual number of vessels. The muscular bundles of the ciliary muscle are obliquely shaded.

(ophthalmia), and will usually be accompanied
by muco-purulent or purulent discharge (2) A
zone of pink congestion surrounding the cornea,
and formed by small, straight, parallel vessels
closely set, radiating from the cornea, and not
moving with the conjunctiva (anterior ciliary *arterial* twigs), points to irritation or inflammation of
cornea (foreign body, ulcer, diffuse keratitis), or to
iritis. A similarly placed zone of dark or dusky
colour, and finely reticulated (episcleral *venous
plexus*) often points to glaucoma, but is frequent in
other conditions, especially in old people. Congestion in the same region, more deeply seated, and of a
peculiar lilac tint, especially if unequal in different
parts of the zone, shows cyclitis. (3) Congestion in
the same zone, and also composed of small vessels,
but superficially placed, bright red, and often encroaching a little on the cornea (*anterior conjunctival
vessels* and *loop-plexus of cornea*), usually shows a
tendency to the relapsing forms of superficial corneal
ulceration (phlyctenular affections). It is only necessary to add that in the severe forms of any acute disease of the front of the eye these types of congestion
are often mixed and but imperfectly distinguishable,
but that much information is often derived from
attention to the leading forms described. Localised
or fasciculated congestion generally points to phlyctenular disease.

(8.) Note the *colour of the iris*, and compare it
with that of the fellow eye. In some persons the
irides, although healthy, are of different colours in
the two eyes, one blue or grey, the other brown or
greenish ; and sometimes one iris shows large patches
of lighter or darker colour than its fellow (piebald).
But if the iris of an inflamed eye is greenish while its
fellow is blue we should suspect iritis, and if the iris
of a defective eye be different from its fellow signs of
former iritis should be sought.

(9.) *The pupils* are to be examined as to (1)

equality, (2) size in ordinary light, (3) mobility, (4) shape. Under mobility we should note first the behaviour of each pupil separately to light and shade, with both eyes open, and again with each alternately closed; second, the behaviour of both pupils when shaded and lighted together; third, their behaviour in relation to accommodation and convergence; and, lastly, the effect of atropine in cases needing it. The pupils in health should be round and equal, and should be opposite to, or a little to the inner side of, the centre of the cornea. When one eye is shaded its pupil should dilate considerably, and the other to a less extent, and when again exposed it should contract briskly. The pupils are often large and inactive in amaurotic patients, in glaucoma, and in paralysis of the third nerve, but the last is seldom symmetrical. They may be too large but still active in myopia and in conditions of defective nerve tone. Wide dilatation of one or both pupils, with dimness of sight of a few days' duration, and without ophthalmoscopic signs of disease, is usually traceable to the use of atropine or belladonna either by accident or on purpose. The pupils are smaller and less active in old age, but unusual smallness and inactivity at any age should excite suspicion of spinal or cerebral disease, unless posterior synechiæ are present (past iritis). When very small the pupil is seldom quite circular; and, again, when dilated from old disease it is often oval, and usually in the vertical direction; it is rare to see a pupil oval horizontally. The pupil may remain motionless when shaded or brightly lighted either from iritic adhesions, paralysis of its muscular fibres, or glaucoma; and, lastly, from loss of reflex stimulus from the retina of its own eye if this be blind (disease of optic nerve, detached retina, &c.), the iris itself being healthy. In the last-named case the pupil of the blind eye will act concomitantly with that of the other (healthy) eye, since stimulation of one retina excites reflex action of both irides. The pupils should dilate when the ciliary muscle and internal recti are relaxed (gazing into the distance), and contract during accommodation and convergence (looking at an object distant a few inches). The dilatation effected by atropine is often less in old than in young people. In comparing the two pupils take care that both are equally well lit; marked inequality is rare, excepting from disease or from widely different refraction in the two eyes. When very active pupils are suddenly exposed after being shaded they often oscillate for a few seconds before settling, and finally remain a little larger than at the first moment of exposure. The pupils are smaller whenever the iris is congested, whether this be a merely local condition (*e. g.* in abrasion of cornea), or form part of a more general congestion, as in typhus fever* and in plethoric conditions, or be caused by venous obstruction, as in mitral regurgitation and bronchitis. They are large in anæmia, and in cases where the systemic arteries are badly filled, such as aortic insufficiency,† and during rigors.

(10.) *The field of vision* is the surface from which, at a given distance, light reaches the retina, the eye being at rest (Fig. 3). If each part of the field is equidistant from the part of the retina to which it corresponds, the field will form the inner surface of a hemisphere turned towards the eye; it may, however, be projected on to a flat surface, and for many clinical purposes this is quite accurate enough. For roughly testing the field, *e. g.* in a case of chronic glaucoma, or of atrophy of optic nerve, or of hemiopia,

---

* The small pupil of typhus and the frequently large pupil of typhoid are ascribed by Murchison to the differences in the vascularity of the iris (as a part of the whole eyeball) in the two diseases. 'Continued Fevers,' 541.

† See an article on 'The Indications afforded by the Pupil,' 'Medical Examiner,' March 2, 1879. Argyll Robertson, however, doubts the influence of changes of vascularity of the iris on the size of the pupil. 'Edin. Med. Jour.,' 1869, 703.

the following is generally enough. Place the patient with his back to the window; let him cover one eye, and fix his gaze steadily upon the centre of your face or nose at a distance of 18″ or 2′. Then hold up your hands with the fingers spread out in a plane with your face, and ascertain the greatest distance at which they are visible in various directions —up, down, in, out, and diagonally. Of course it is essential that the patient should look steadily at one's face, and not allow his eye to wander after the moving fingers.

A more accurate method is to make the patient gaze, with one eye closed, at a white mark (the fixation spot) on a large black board (a distance of 1′ is generally chosen), and to move a piece of white chalk set in a long black handle from various parts of the periphery towards the fixation spot, until the patient exclaims that he sees something white. If a mark be made on the board at each of about eight such peripheral points, a line joining them will give with fair accuracy the boundary of the visual field.

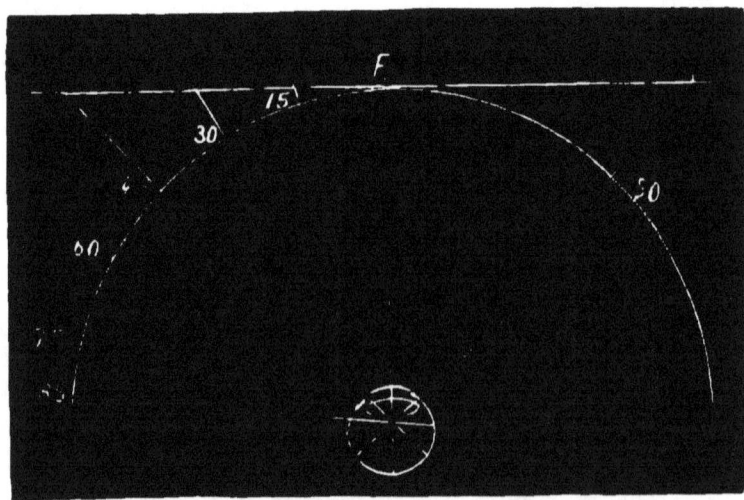

FIG. 3.—Field of vision, projected up to 45° on to a flat surface two feet square. F, fixation spot.

By having a flat board 2' square, divided into circles at distances corresponding to equidistant spaces on the true (hemispherical) field (Fig. 3), greater precision is attained up to about 45°; but beyond that distance accurate maps of the field can be made only by means of an accurate instrument, the perimeter, which consists essentially of an arc marked in degrees, and moveable around a central pivot on which the patient fixes his gaze. The visual field in health is not circular, but forms an oval with the small end upwards and inwards. From the fixation point it extends 90° or more in the outward direction, but only about 55° in the inward, inward-downward, and inward-upward, and 45° in the upward meridian.

(11.) *Numeration of spectacle lenses.*—Two systems of numbering are in use. In the first, which has till lately been universal, the unit taken was a lens of 1″ focal length. As all the lenses in ordinary use are weaker than this, their relative strengths can be expressed only by using fractions. Thus, a lens of 2″ focus is half as strong as the unit, and is therefore expressed as $\frac{1}{2}$; a lens of 10″ focus is $\frac{1}{10}$; of 20″ focus $\frac{1}{20}$; and so on. The inconvenience of using fractions is considerable, and further, as the length of the inch is not the same in all countries, a glass of the same *number* has a somewhat different focal length according as it is made by the Paris, English, or German inch. The *inch scale* for numbering lenses is, therefore, giving place to the *metrical scale.* In this scale the unit is a lens of 1 metre focal length, and for convenience it is called a *Dioptric* (D). A lens of double this strength, or half a metre focal length, is 2 dioptrics (2 D), a lens of ten times the strength, or one tenth of a metre focus, is 10 D, and so on. The weakest lenses are ·25, ·5, and ·75 D, and in the higher powers intermediate numbers differing by ·5 or ·25 D are also introduced. The universal adoption of this system will lead to much greater uniformity in the manufacture of spectacles, and to the avoidance of some trouble in adding and subtracting fractions. If it is desired to convert one system into the other, this can be done provided that we know what inch was used in making the lens whose equivalent is required in Dioptrics. The metre is equal to about 37 Paris and 39 English or German inches; a lens of 36 Paris inches (No. 36 or $\frac{1}{36}$ old scale), or of 40 English or German inches (No. 40 or $\frac{1}{40}$), is very nearly the equivalent of 1 D. A lens of 6 Paris inches ($\frac{1}{6} = \frac{6}{36}$ will therefore be equal to 6 D; a lens of 18 Paris inches ($\frac{1}{18} = \frac{2}{36}$) = 2 D, &c. A lens of 4 D = $\frac{4}{36}$ = $\frac{1}{9}$, or a lens of 9 Paris inches, &c.

(12.) *Testing the acuteness of sight.*—By acuteness of sight (V. or S.) is meant the power of distinguishing *form*, and the standard taken as normal is the power of distinguishing square letters which subtend an angle of 5 minutes, and whose limbs, which are of uniform thickness, are seen separately under an angle of 1 minute (Snellen's Test Types). As commonly used the term refers only to the centre of the visual field, the peripheral parts of the retina having only a very imperfect power of distinguishing form and size. Rays forming so small an angle are very nearly parallel, and are united on the retina without any effort of accommodation. These types are made of various sizes, each size being numbered according to the distance (either in feet or metres), at which it subtends a visual angle of 5 minutes. Thus, No. XX subtends this angle at 20', No. X at 10', No. II at 2'. Numerically, acuteness of vision is expressed by a fraction, of which the denominator is the number of the type, and the numerator the greatest distance at which it can be read ; if No. XX is read at 20' V. = $\frac{20}{20}$, or 1 or unity, *i.e.* normal ; if only No. L can be read at 20' V = $\frac{20}{50}$; if only CC, then V = $\frac{20}{200}$. Any distance more than about 6' may be selected for this test, *i.e.* No. VI read at 6', or No. X at 10',

generally show the same acuteness as XX read at
20'. But at shorter distances the accommodation
comes into play, and other fallacies are introduced,
and hence $\frac{2}{2}$ is not the same as $\frac{20}{60}$. It is, therefore, best to record the fractions unreduced so that
the distance at which the test was used may be
known. For testing near vision, indeed, Snellen's
types are thought by many to be practically inferior
to those of Jaeger and others, in which the letters have
the form and proportions found in ordinary type.

(13.) *Accommodation* (A.) is tested clinically by
measuring the nearest point (*punctum proximum*, or
p.), and the furthest (*punctum remotum*, or *r.*) of distinct vision for the smallest readable type (Snellen's
or Jaeger's $1\frac{1}{2}$ or 1). The *amplitude* or *range* of accommodation is expressed by the number of the convex
lens which can be substituted for A, and this can
always be found by the formula $\frac{1}{p} - \frac{1}{r}$. Thus, No. $1\frac{1}{2}$
of Snellen's types is read at 5″ (p.), and at 18″ (r.);
in the latter position it is seen under an angle of 5
minutes, *i.e.* without effort of accommodation, and
may be considered as infinitely distant; hence
$\frac{1}{p} - \frac{1}{r} = \frac{1}{5} - \frac{1}{\infty} = \frac{1}{5}$; in other words, the accommodation is equal to a lens of 5″ focal length.
The *region* of accommodation is the space in which
it is available; thus, in a case of myopia p. $= 3″$,
r. $= 7\frac{1}{2}″$; accommodation is exerted within that space
because beyond $7\frac{1}{2}″$ no objects are seen distinctly, but
its *amplitude* is as great as in the former case, since $\frac{1}{3}$
$- \frac{1}{7\frac{1}{2}} = \frac{1}{5}$.

Relative accommodation. With every degree of convergence of the optic axes a certain definite quantity
of accommodation is involuntarily associated; this
quantity, however, is capable of increase and of decrease within certain limits without alteration of convergence. Thus, although in converging the optic
axis to a point 12″ off, we also involuntarily accom-
modate our crystalline lens so as to gain clear vision
at that distance (*i.e.* we use accommodation equal to

a lens of 12″ focal length), it is possible on the one hand to use somewhat less accommodation, and on the other to exert considerably more than $\frac{1}{12}$, without changing the position of the eyes. The part which, being held in reserve, can be used without increasing the convergence of the eyes, is the *positive*, and the part which, being in use, can be relaxed without altering the convergence, is the *negative* part of the relative accommodation. In order to maintain accommodation for long at a given distance, the positive part of the relative accommodation must be tolerably great.

(14.) *The apparent size* of an object depends, in the first place, on the size of its *retinal image*. In Fig. 4,

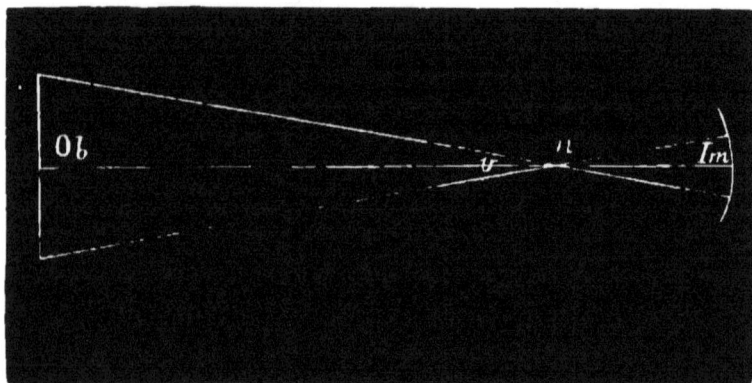

Fig. 4.—Visual angle.

if *ob* is an object, *n*, the optical centre of the eye at which the axial rays cross one another and *Im*, the retinal image, it is evident that the size of *Im* will vary with the size of the angle, *v*, formed by lines drawn from the extremities of the object to *n*. This is the *visual angle*. It is clear that a smaller object placed nearer to the eye or a larger one placed further off may subtend the same angle as *ob*, and therefore have a retinal image of the same size. There are, however, other factors contributing to our

estimate of the size of objects, especially contrast of
size and shade, estimation of distance, and effort of
accommodation.

A white object on a black ground looks larger than
a black object of the same size on a white ground.
The further off an object appears to be, the larger
does it seem to our judgment to be. The greater
the accommodative *effort* used, independently of the
distance of the object, the smaller does it appear to
be ; thus, patients whose eyes are partly under the
influence of atropine, and presbyopic persons whose
glasses are too weak, complain that near objects
looked at intently presently get much smaller ; whilst
when one eye is under the action of eserine (causing
spasm of the accommodation) objects appear larger
than they do at the same distance to the other eye.
Prisms with their bases towards the temples seem to
diminish objects seen through them by necessitating
excessive convergence of the eyes.   The reverse
occurs if they be turned with their bases towards the
nose.

(15.) *Colour perception* is best examined by testing
the power of discriminating between various colours
without naming them. The best test-objects are a
series of skeins of coloured wool, or, for pocket use,
smaller strips of coloured paper, or, better, of
coloured stuffs.* A colour-blind person will at once
expose his defect by placing side by side as similar,
certain colours, usually mixed tints, which to the
normal eye appear quite different. The set of wools
which I use has been selected as nearly as possible
according to the directions given by Dr Holmgren,
of Upsala, in his recent work on the subject.† In
acquired colour-blindness (from atrophy of the optic
nerves), the patient, if well trained in colours,

---

* A set of coloured stuffs of convenient size for the pocket
can be obtained of Mr Hawksley, 300, Oxford Street.
† ' De la Cécité des Couleurs,' &c., 1877.

may be asked to name the colours, and his defect
will generally in this way be correctly found;
but in congenital colour-blindness the confusion
test, without naming the colours, is far safer because,
in the first place, such a patient has often learnt
to distinguish correctly between many common
coloured objects by the differences of *shade* (*i. e.*
differences in the quantity of white light which they
reflect), and hence may escape detection unless tested
with many different colours, some of which containing equal quantities of white will look to him
exactly alike; secondly, though such persons often
use the names for colours freely, the words do not
convey to them the same meaning as to others, and
hopeless confusion results from an examination so
made.

The spectroscope may also be used for testing
colour-blindness, but I have not found it of much
clinical use.

(16.) *Prisms* are occasionally useful in examining
feigning patients, for the correction of diplopia from
slight strabismus, for estimating the strength of the
internal and external recti, and other purposes.

When light passes through a prism it is bent
towards its base (thick end), and on entering the eye
appears to have come in this altered direction.
Thus, if a prism be held before the right eye, with its
base towards the nose, rays from an object in the
middle line after traversing the glass will be deviated
inwards, and will seem to the right eye to have come
from a point somewhere to the right of the actual
object. If both eyes are open, the first effect will be
double vision, because the prism throws the image
in the right eye to the inner side of the yellow
spot; but the eye will immediately make a com-
pensating movement outwards until the image is
again on the yellow spot and single vision regained.
If the prism used be so strong that the greatest
effort of the external rectus cannot move the eye

sufficiently outwards, the diplopia will remain. The internal rectus can overcome a much stronger prism (base outwards) than the external (base inwards), whilst the superior and inferior recti are still less able to cause compensatory movements (base downwards or upwards); and in endeavouring to cause diplopia in an examination for malingering it is best to use a rather weak prism (about 6° or 8°) with its base up or down.

CHAPTER III

EXAMINATION OF THE EYE BY ARTIFICIAL LIGHT

THIS includes (1) examination by focal or oblique light; (2) examination by the ophthalmoscope.

(1.) Examination by *focal* or *oblique illumination* is meant the inspection of the anterior parts of the eye with the light of a lamp concentrated by means of a convex lens. It is used for the examination of opacities of the cornea, changes in the appearance of the iris, especially as regards its vascularity, irregularities of the pupil and pigment and membrane in its area (iritic adhesions), and, lastly, opacities of the lens. Such an examination is to be made by routine in every case before using the ophthalmoscope. We require a somewhat darkened room, a convex lens of two or three inches focal length (one of the large ophthalmoscope lenses), and a bright naked lamp-flame.

The patient is seated with his face towards the light, which is at about 2' distance. The lens, held between the finger and thumb, is used like a burning-glass, being placed at about its own focal length from the patient's cornea and in the line of the light, so as to throw a bright pencil of light on the front of the eye at an angle with the observer's line of sight. Thus all the superficial media and structures of the eye can be successfully examined under strong illumination, the lens being brought rather nearer or removed a little further, so that its focus falls as desired on the cornea, the iris or the anterior or posterior surface of the crystalline lens. By varying the position of the light and of the patient's eye, making him look up, down, and to each side, we can thoroughly examine all parts of the corneal surface, of the iris, of the pupillary area (*i. e.* the anterior capsule of the lens), and of the lens substance. By throwing the light at a very acute angle on the cornea or lens opacities become much more visible than if it be thrown almost perpendicularly.

For complete exploration of all parts of the crystalline lens the pupil must be dilated with atropine, but careful examination without atropine will generally enable us to detect opacities lying in or near the axis of the lens even if quite deeply seated. In examining the posterior pole of the lens the light must be thrown almost perpendicularly into the pupil, and the observer must place his eye as nearly in the same direction as is possible without intercepting the incident light.

Opacities of the cornea and lens appear white or whitish yellow by focal light. Tumours and large opacities in the vitreous, hæmorrhagic or other, may be seen by this method if seated close behind the lens. Minute foreign bodies in the cornea will often be seen by focal light when invisible, because covered by hazy epithelium, in daylight. Lastly, by using a second lens in the other hand the parts illuminated can be magnified, and much additional information gained. The habitual use of a magnifying-glass for inspecting the front of the eye cannot be too much recommended.

(2.) *Ophthalmoscopic examination*

The ophthalmoscope enables us to see the parts of the eye behind the crystalline lens by making the observer's eye virtually the object by which the observed eye is lighted up. Rays of light entering the pupil in a given direction are partly reflected back by the choroid and retina, and on emerging

from the pupil again take the same or very nearly
the same course that they had on entering. Hence
the eye of the observer if so placed as to receive these
returning rays must also be so placed as to cut off
the entering rays; as, therefore, no light will enter in
this direction, none will return to the observer's eye.
Hence the pupil generally looks black. Although
with a large pupil it is possible for the observer to
receive some of the returning rays without intercepting the entering light, and in this way to see the
pupil of a fiery red instead of black; still for any
useful examination the observer's eye must, as
already stated, be in the central path of the entering
(and emerging) rays. This end is gained by looking
through a small hole in a mirror, by which light is
reflected into the patient's pupil, and this perforated
mirror is the ophthalmoscope.

In order to see the parts at the back of the eye
distinctly we may proceed in one of two ways

A. *The indirect method* of examination, by which a
clear, real, inverted image of the fundus somewhat
magnified, is formed in the air between the patient
and the observer.

The following simple experiment will show how
this is effected :—Take two convex lenses of about
2″ focal length each. (1) Hold one in the left hand,
and look through it with one eye closed at this print,
holding it as far as possible from the page; the
print will, of course, be magnified, and its position
unchanged; (2) take the second lens in the right
hand, and, removing your head a few inches back,
hold it at about its own focal length in front of the
first ; you will then see an inverted image of the
print slightly magnified, though less so than when
seen through the first lens. You will notice that in
order to see this inverted image clearly you have to
make an effort, and that you cannot see this image
and the print on the page clearly at the same

moment; this is because the inverted image (*im*, Fig. 5) is in the air between the eye and the second lens, and more accommodation is necessary for seeing it clearly than for the object (*ob*). The fundus of the eye thus seen is magnified about four diameters, if the eye be normal. It is larger in

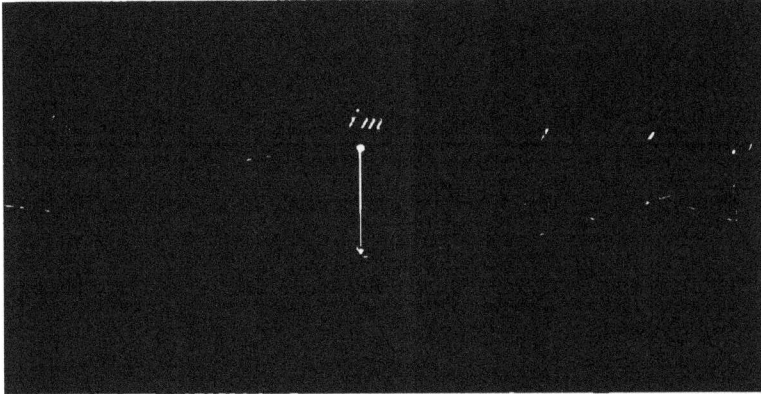

FIG. 5.—*ob* is the object. *a*. The first lens. *l*. The second lens. *im*. The magnified inverted image viewed by the eye, *obs*.

H and smaller in M. Lastly, if the observer's head be moved slightly from side to side the image will appear to move in the opposite direction.

B. *The direct method* of examination, by which a virtual, erect image is seen more magnified than in the former method and behind the patient's eye.

The conditions are the same as those under which a magnified image of any object is seen through a convex lens, as in the following experiment :—(1) Take a convex lens of, say, $2\frac{1}{2}''$ focal length, and hold it at any distance from this page not greater than $2\frac{1}{2}''$, and place your eye close to the lens. The print will be magnified and seen in its true position, *i. e.* "erect." The enlargement will be greater the greater the distance of the lens from the page up to $2\frac{1}{2}''$. Any further increase of distance will make it impossible to see the print clearly. The image is a "virtual" one, because it is the image which would be formed if the rays which enter the eye in a diverging direction could be prolonged backwards until they met behind the lens (Fig. 7). If the lens is placed just at its focal length from the paper the image will be seen clearly only during complete relaxation of the accommodation; if it is nearer to the page, either the accommodation must be used in proportion to the distance, or the observer must withdraw his head to a distance from the lens. Again, if, keeping the lens quite still, the observer withdraw his head, the image will appear to increase in size, but the field of view will be lessened, and this change will be greater the nearer the lens is to its focal distance from the paper; if it is almost exactly at its principal focal distance, only a very small part of the print will be seen when the head is withdrawn, but this part will be very highly magnified. Lastly, if the head be moved a little from side to side the image will appear to move in the same direction.

When the eye is properly formed the retina is placed almost exactly at the principal focus of the lens-system of the eye,* so that when the accommodation is fully relaxed, i. e. when the eye is adjusted for rays of light from distant objects which are practically parallel, a clear image of such object is formed on the layer of rods and cones of the retina.

A clear image of the fundus can be obtained in such an eye just as in the second experiment above described, when the distance of the lens from the paper was equal to or less than its focal length ; the conditions being that the eyes, both of the patient and observer, must be adjusted for infinite distance, i. e. for parallel rays ; in other words, that the accommodation of both must be relaxed.

In order *to use the ophthalmoscope*† it is first

---

* About 22·25 m.m. from the front of the cornea and 15 mm. from the posterior surface of the crystalline lens.

† Liebreich's "small" ophthalmoscope is the most convenient

necessary to learn to manage the mirror and light.
(1.) Seat the patient in a darkened room and place
a lamp with a large steady flame from an argand
burner, on a level with his eyes, a few inches from
his head, and about in a line with his ear. The
lamp may be on either side, but is usually placed
on his left, and it is better to keep to the same,
side until practice has given steadiness to the various
little combined movements which are necessary. (2.)
Sit down in front of the patient with his face front-
ing yours feature to feature. It will be found most
convenient for the observer's face to be a little
higher than that of the patient. (3.) Take the
mirror of the ophthalmoscope (without any lens in
the clip behind, and without the large lens) in your
left hand for examining the patient's left eye (and
vice versâ for his right eye), hold it, mirror towards
the patient, close to your own eye, and with the sight-
tube placed so that (with your other eye closed) you
see the patient through it. Now rotate the mirror
slightly towards the lamp until the light reflected
from the flame is thrown on to the patient's eye, and
open your other eye. (4.) You will so far have seen
nothing except the front of the eye, unless the patient's eye is under atropine. He will be sure to
look at the centre of the mirror, and his pupil
will look either black or very dull red. (5.)
Now tell him to look steadily a little to one side
into vacancy, or at an object on the other side
of the room. The pupillary area will now become
red ; bright fiery red if the pupil is rather large ; a
duller red if the pupil is very small or the patient be

of the inexpensive forms for ordinary use. The sight-hole
should be 2 mm. in diameter and its outline smooth, and it need
not be perforated. The two large lenses serve very well for
trying the experiments described in the text. But when expense
is not a great object the student will be well repaid by at once
buying one of the " refraction ophthalmoscopes " mentioned at
p. 40.

of dark complexion. In one position when the eye under examination looks a little inwards the red will change to a yellowish or whitish colour, and this indicates the position of the optic disc. (6.) Learn to keep the light steadily on the pupil during slow movements backwards and forwards and from side to side (taking care that the patient keeps his eye all the time in the same position, and does not follow the movements of the mirror) ; the test of its steadiness will be that the pupil remains of a good red colour in all positions. Up to this point the examination may be made without atropine; and so far no details of the fundus will have been seen, only a uniform red glare, unless the patient be either myopic or considerably hypermetropic.

In order to see the details of the fundus it is best to begin by learning *the indirect method* (Fig. 6), for though rather less easy than the direct it is more generally useful.

Having learned to keep the light reflected steadily into the patient's pupil, take the mirror without any lens in the clip behind it (unless you are either hypermetropic or myopic, in which case you should either wear the glasses you commonly use for reading, or place one of the small lenses supplied with the ophthalmoscope, and of about the same focal length, in the clip behind the mirror) in one hand, and one of the large convex ocular lenses in the other. Always, if possible, have the pupil dilated with atropine, for by this means you learn to see the fundus much more quickly and easily. Apply the mirror with your right hand to your right eye, holding the lens in your left hand, for the patient's right eye, and reverse everything for his left eye ; but the position of the light need not be changed. The hand which carries the lens should be steadied by resting the little or ring finger against the patient's eyebrow or temple.

OPHTHALMIC EXAMINATION

It is best to begin by looking for the optic disc, which is one of the most important and easily seen parts. To bring it into view the patient must look a little inwards, with the eye under examination, *e. g.* if his right eye is under examination, he must direct it to the observer's right ear, or look at the little finger of his mirror hand. Take care that the patient turns his eye, not his head, in the required direction. The lens should be held about 2″—3″, and the observer and mirror be about 18″ from the patient's eye; the image of the fundus being formed in the air about 2½″ in front of the large lens will thus be situated about 12″ from the observer.

The bright red glare of the fundus will be obvious enough; but most beginners find more or less difficulty in adjusting the distance of their head and accommodation, so as to see the aërial image clearly. The head must be slowly moved a little further or nearer to the patient, and at the same time an attempt made to adjust the eyes (for they should both be open) for a point between the observer and the lens. Several sittings are sometimes necessary before the image of the optic disc, or even of any retinal vessels, can be clearly seen.

The optic disc—ending of the optic nerve in the eye above the lamina cribrosa, optic papilla (Figs. 8 and 9)—is seen as a round object, of much lighter colour than the fiery red of the surrounding fundus, and from its centre numerous blood-vessels radiate, chiefly in an upward and downward direction. As soon as it can be easily seen the student must pass on to the study of the most important details of the disc itself and of the other parts of the fundus, some of which will be given here and others will be found in the chapters on the diseases of the choroid and retina, and on the errors of refraction.

The disc, as a whole, is of a greyish pink with admixture of yellow. It is nearly circular, but seldom

perfectly so, being often apparently oval or slightly
irregular. Two different colours are noticeable—a
central spot or patch, whiter than the rest, and sometimes stippled with several greyish dots (*lamina
cribrosa*), and into which most of the blood-vessels
dip; and a surrounding part of pink or greyish pink,
In many eyes the apparent boundary of the disc is
formed by a narrow line of lighter colour, which represents the border of the sclerotic. The blood-vessels
consist of several large trunks and a varying number
of small twigs; the large trunks emerge from the
central white part of the disc, and often bifurcate
once or twice on its area; the small twigs emerge
from various parts of the disc, or are branches of the
large trunks.

Variations are numerous. The colour of the disc
appears paler or darker according to the colour of
the surrounding choroid, the brightness of the light
used, and the patient's age and state of health. A
curved line of dark pigment often bounds a part of the
circumference of the disc and has no pathological
meaning. The central white patch varies greatly
in size, position, and distinctness; it may be so
small as hardly to be perceptible, or very large; may
shade off gradually or be abruptly defined; may be
central or eccentric; when large it generally shows
a greyish stippling or mottling. This white patch
represents a depression of corresponding position
and size, the *physiological cup* or *pit* (compare
Figs. 8 and 9) formed by the nerve-fibres radiating from the centre of the disc on all sides towards the retina, like the tentacles of an open sea
anemone, and through it the chief blood-vessels
pass on their way between the nerve and the retina.
This depression is generally shaped like a funnel or
a dimple, with gradually sloping sides (Fig. 9);
but sometimes the sides are steep, or even overhanging; in other eyes it is wide or shallow, and
enlarged towards the outer side of the disc. The

3

physiological pit is whiter than the rest of the disc, because the greyish-pink nerve-fibres are absent at this part, and we can therefore see down to the opaque, white, fibrous tissue which, under the name of lamina cribrosa, forms the floor of the whole disc (Fig. 9). Its frequently stippled appearance is caused by the holes in this lamina, through which the bundles of nerve-fibres pass on their way to the retina, the holes appearing darker because filled by non-medullated nerve-fibres, which reflect but little light.

Appearances of the other parts of the fundus.— The groundwork is of a bright fiery red (the choroid), which in average eyes is pretty uniform, but in persons of very light or very dark complexion shows a closely-set tortuous pattern of red bands (vessels) separated by interspaces either of lighter or of darker colour (for further details *see* Diseases of Choroid).

Upon this red ground the vessels of the retina are seen dividing and subdividing dichotomously. It will be noticed that the principal trunks pass nearly upwards and downwards, but that no large branches go to the part apparently inwards from the disc ; that the whole number of visible retinal vessels is comparatively small, large spaces intervening between them ; that they become progressively smaller as they recede from the optic disc ; and that they never anastomose with each other. Special attention must be given to the part apparently to the inner (nasal) side of the optic disc (really to its outer, temporal side) which is the region of most accurate vision, the yellow spot. This region is skirted by large vessels from which numerous twigs are given off to it, but no vessels cross its centre (*see* Diseases of Retina). The yellow spot, *y. s.* (macula lutea, or shortly " macula ") is seen when the patient looks straight at the ophthalmoscope ; it will be noticed that the choroidal red is darker at this part, and that no retinal vessels pass across

its centre, but that numerous fine twigs radiate
to and from it. In many eyes nothing but these
indefinite characters mark the yellow spot. But in
some, especially in dark eyes and young patients,
a very minute bright dot occupies the very centre,
and is encircled by an ill-bounded dark area, round
which again a characteristic shifting white halo is
seen. The minute dot is the *fovea centralis*, the
thinnest part of the retina. The peripheral parts of
the fundus are explored by telling the patient to look
successively up, down, and to each side without moving his head. To see the extreme periphery the
observer must move his head as well as the patient
his eye. Towards the periphery the choroidal trunk
vessels are often plainly visible when none were
distinguishable at the more central parts.

The vessels of the retina are easily distinguished
from those of the choroid by their course and mode
of branching, and by the small size of all except the
main trunks, but especially by their greater sharpness of outline and clearness of tint, and by the
presence of a light streak along the centre of each,*
which gives them an appearance of roundness, very
different from the flat band-like look of the choroidal
vessels. They are easily divisible into two sets—a
darker, larger, somewhat tortuous set—the veins;
and a lighter, brighter red, smaller, and usually
straighter set—the arteries, the diameter of corresponding branches being about as 3 to 2. They run
pretty accurately in pairs.

The indirect method of examination is most generally useful, because it gives a large field of view,
under a comparatively low magnifying power (about
three to five diameters). The general characters and
distribution of any morbid changes are better appre-
ciated by this means than if we begin with the
direct method, in which the field of view is smaller
and the magnifying power much greater. It has

* Not accurately shown in Fig. 8.

also the great advantage of being equally applicable in all states of refraction in the patient, whereas the erect image can be obtained in myopia only by the aid of a suitable concave lens placed behind the mirror, and found experimentally. In the inverted image the inversion is such that what appears to be upper is lower, and what appears to be on the R. is really on the L.

The *direct method, i. e.* examination by the mirror alone, or with the addition of a lens in the clip behind it, but without the intervention of the large lens.

By this method the parts are seen in their true position (Fig. 7), and it is hence often called examination of the "erect" or "upright" image; as will be seen, however, the latter terms are not strictly convertible with "direct examination." It is used (1) to detect opacities in the vitreous humour and detachments of the retina. (2) To ascertain the condition of the patient's refraction, *i. e.* the relation of his retina to the focal distance of his lens system. (3) For the minute examination of the fundus (highly magnified, virtual, erect image, Fig. 8).

(1.) To examine the vitreous humour. The patient is to move his eye freely in different directions whilst the light is reflected into the eye from a distance of a foot or more (for details see diseases of vitreous). Detachments of the retina are seen in the same way. Opacities in the vitreous and folds of detached retina are seen in the erect position, and usually at a considerable distance from the eye, because they are situated far within the focal length of the lens system, and are therefore seen under the conditions mentioned at p. 28.

(2.) To ascertain the kind of refraction. If when using the mirror alone at a good distance (2' or 18") from the patient's eye we see some of the retinal vessels clearly and easily, the eye is either myopic

or hypermetropic.   If, when the observer's head
is moved slightly from side to side, the vessels
seem to move in the same direction, the image seen
is a virtual one, and the eye is hypermetropic;
but if the vessels seem to move in the contrary
direction the eye is myopic.   This image in
myopia is, indeed, formed and seen in the same way
as the inverted image seen by the " indirect " method
of examination, but excepting in the highest degrees
of myopia it is too large and too far from the
patient to be available for detailed examination.   In
the low degrees of myopia it is formed so far in
front of the patient's eye that it will be visible
only when the observer is distant perhaps 3' or 4';
whilst if the eye is of normal refraction or slightly
hypermetropic the erect image will not be easily
seen at so great a distance as 18" or 2' (compare p.
27).

The above tests only reveal qualitatively the presence of either myopia or hypermetropia, but by a
modification of the method, the exact quantity of any
error of refraction, e. g. hypermetropia, can be determined with great accuracy (*determination of the refraction by the ophthalmoscope*).   With normal refraction, *i. e.* when with accommodation relaxed parallel
rays are focussed on the retina, the erect image can
be seen only if the observer be very close to the
patient and also completely relax his own accommodation.   In the experiment described at p. 27
with a convex lens, when the head is withdrawn from
the lens the magnifying power appears to increase,
and the field of view and illumination rapidly
diminish.   The same occurs with the eye, but in a
much greater degree, and hence in emmetropia no
useful view can be gained at any considerable distance from the eye.

In hypermetropia, where the retina is within the
focus of the lens system, the erect image is seen
close to the patient's eye only by an effort of

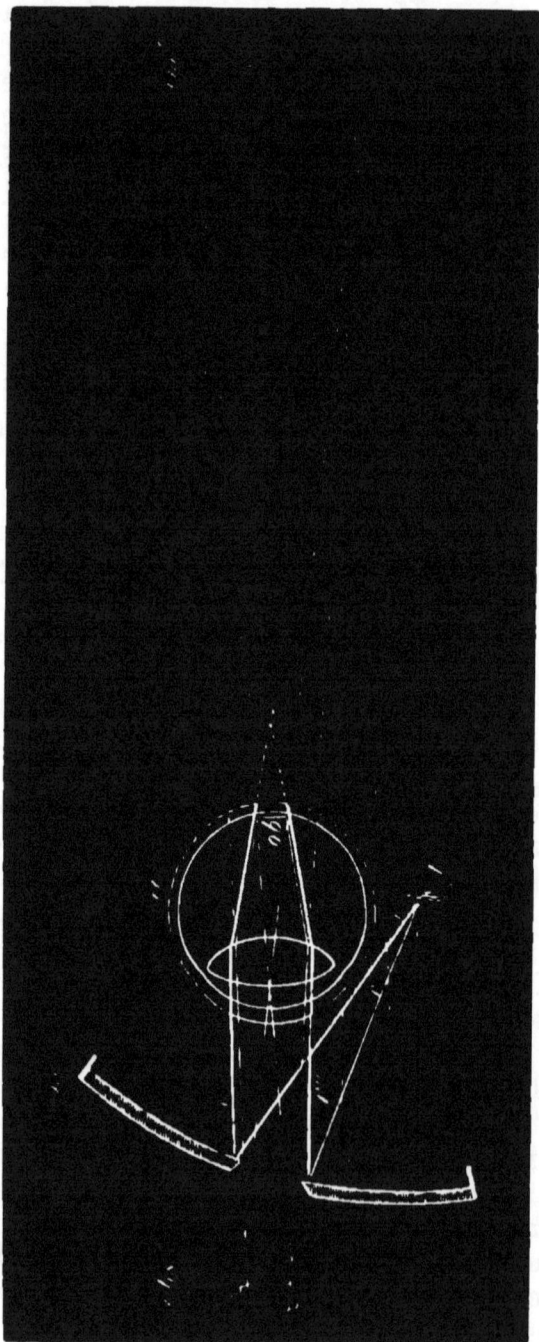

FIG. 7.—Examination of virtual erect image. Lettering as in Fig. 6. The rays r' r'' entering the eye divergent would be focussed behind the retina, as at f, and hence illuminate the fundus diffusely. The returning pencils (thin lines) are parallel or divergent (according as the eye is normal or hypermetropic) on leaving the eye, and appear to the observer to proceed from a highly magnified erect image, im', behind the eye. It is seen that only the lamp-rays which strike close to the sight-hole are available.

accommodation in the observer, just as in the same experiment when the lens was within its focal length from the page. And as in that experiment the print was also seen easily, even when the head was withdrawn, so in hypermetropia the erect image is seen at a distance as well as close to the patient.

If now the observer, instead of increasing the convexity of his crystalline, place a convex lens of equivalent power behind his ophthalmoscope mirror, this lens will be a measure of the patient's hypermetropia, *i.e.* it will be the lens which, when the patient's accommodation is in abeyance, will be needed to bring parallel rays to a focus on his retina. If a higher lens be used, the result will be the same as when in the experiment the convex lens was removed beyond its focal length from the print; the fundus will be more or less blurred.

Hence to measure the hypermetropia (1) the accommodation of both patient and observer must be fully relaxed (usually by atropine in the patient and by voluntary effort in the surgeon). (2) The observer must go as close as possible to the patient; (3) he must then place convex lenses behind his mirror, beginning at the weakest and increasing the strength till the highest is reached with which the details of the optic disc can be seen with perfect clearness. By practice the distance between the corneæ of the patient and observer can be reduced to about $\frac{1}{5}''$. For this purpose the light must always be on the side under examination, in order to avoid much rotation of the mirror. The right eye must examine the right, and *vice versâ*.

In the same way, though with less accuracy in the high degrees, myopia can be measured by means of concave lenses; the lowest lens with which an erect image is obtained being the measure of the myopia.

Astigmatism may also be measured with considerable accuracy by this method; the refraction being estimated first in one and then in the other of the two chief meridians by means of corresponding retinal vessels (*see* Astigmatism).

This application of the direct method needs much practice, and for convenience the necessary lenses, which are very numerous, are placed in a thin metal disc, which can be revolved behind the mirror so as to bring each lens in succession opposite the sight-hole. There are many forms of these " refraction ophthalmoscopes," varying in minor details of construction.*

(3.) The erect image is very valuable, on account of the high magnifying power, for the examination of the finer details of the fundus in diseases of the choroid, retina, and optic disc. In health the disc by this method is seen less sharply defined because more magnified; both the disc and the retina often show a faint radiating striation (the nerve-fibres); the *lamina cribrosa* is often more brilliantly white; and

Fig. 8.—Ophthalmoscopic appearance of healthy disc as seen in the erect image. Dark vessels, veins; double contoured vessels, arteries.   × 15 diameters (after Jaeger).

* Several very useful forms from thirty shillings upwards are sold by Pillischer, New Bond Street. More expensive ones of American or French design, and more delicate workmanship, are to be had of Krohne and Sesemann, Duke Street, and of Weiss and Co., Strand.

the pigment epithelium can be recognised as a fine uniform dark stippling.

If the refraction is normal or hypermetropic no lens is needed behind the mirror; if myopic, a concave lens must be placed in the clip behind the mirror, of sufficient strength to give a good clear, erect image. The observer must come as near as possible to the patient. Three concaves, −24, −12, and −6, and two convexes, +12 and + 8, are supplied to fit into the clip behind the mirror of Liebreich's

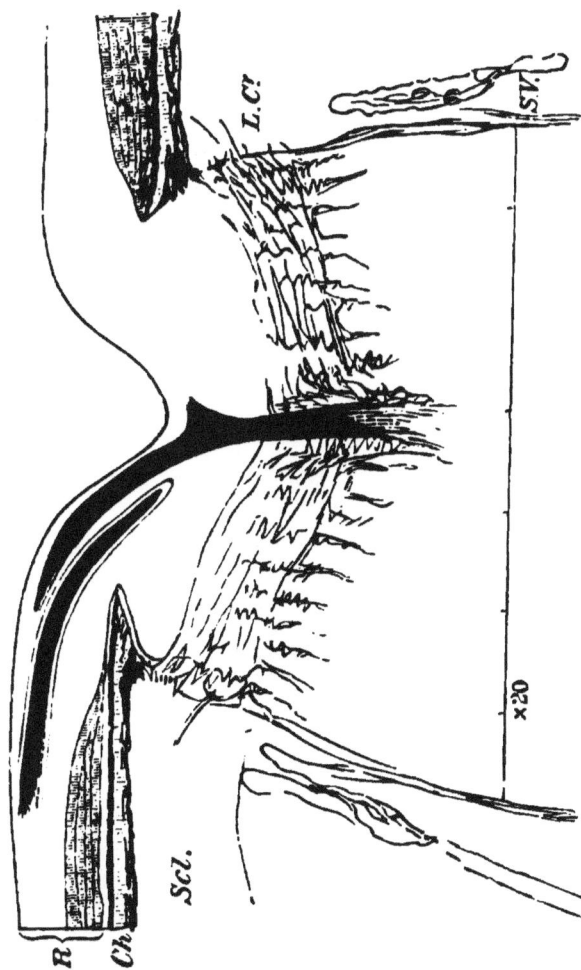

Fig. 9.—Vertical section of healthy optic disc, &c. × about 20. *R.* Retina, outer layer shaded, nerve-fibre layer clear. *Ch.* Choroid. *Scl.* Sclerotic. *L.Cr.* Lamina cribrosa. *S.V.* Subvaginal space between outer and inner sheath of optic nerve. The central vein and a main division of the central artery are seen in the nerve and disc.

ophthalmoscope, and they suffice for the purpose
under consideration in all except very myopic or
extremely hypermetropic patients.

By reference to Fig. 7 it will be seen that only
those rays are useful which strike near the centre
of the mirror, none others entering the patient's
pupil; hence, if the aperture in the mirror be too
large the fundus will not be well lighted, and it
should not be much larger than 2 mm. If much
smaller than that the image has a fictitious clearness
which in certain cases would be misleading.

The details by this method are seen in their real
position; the upper part of the image corresponds to
the upper part of the fundus, and the right to the
right; the image is erect. It is also much more
magnified than by the indirect method, the enlargement in the emmetropic eye being about fourteen
diameters.

PART II

CLINICAL DIVISION

CHAPTER IV

DISEASES OF THE EYELIDS

THE border of the lid which contains the meibomian glands, the follicles of the eyelashes, and certain modified sweat-glands and sebaceous glands, is often the seat of troublesome disease. Being half skin and half mucous membrane, it is more susceptible than the drier and harder skin to irritation by external causes; being a free border, its circulation is terminal, and therefore especially liable to stagnation. Its numerous and deeply-reaching glandular structures, therefore, furnish an apt seat for chronic inflammatory changes.

Ophthalmia tarsi (blepharitis, tinea tarsi, sycosis tarsi) includes all cases in which the border of the eyelid is the seat of subacute or chronic inflammation. There are several types. Generally, neither the skin nor the conjunctiva are so much altered as the parts between them. It may affect both lids or only one, and the whole length or only a part.

In the commonest and most troublesome form the glands and eyelash follicles are the principal seat of the disease. The symptoms are firm thickening and dusky congestion of the border region, with exudation of sticky secretion from its edge, gluing the lashes together into little pencils. In early or very mild cases the lashes are simply overgrown and look coarse. But generally the disease progresses; little excoriations and ulcers covered by scab form along the free border, and often minute pustules appear; the thickening increases; the lashes are loosened, and the parts become so vascular as to bleed freely when they are pulled out. After months or years of varying activity some or all of the hair-follicles are altered in size and direction, or quite obliterated; and stunted, or misplaced, or deficient lashes, are the result; as the thickening gradually disappears, little lines or thin seams of scar are observed just within the edge of the lid, which is often slightly everted by their contraction. The resulting exposure of the marginal conjunctiva, together with the deficiency of lashes, gives a peculiarly raw and bald appearance, to which the term lippitudo is given. Epiphora, from eversion, from tumefaction, or from narrowing of the puncta, is a common result in such bad cases. Often, however, the disease leads to nothing worse than the permanent loss of a certain number of the lashes.

In another type the changes are quite superficial, the patient being liable, perhaps through life, to soreness of the borders of the lids, which are red, and show little crusts and scales at the roots of the lashes, marginal eczema, the growth of the lashes not being interfered with. In such people the eyes look weak or tender; the condition is worse in windy weather, or after long spells of work or exposure to heat or dust. Minute pustules may form around the roots of the lashes.

Ophthalmia tarsi generally begins in childhood, and an attack of measles is the commonest exciting cause. It seldom becomes severe or persistent unless there are both neglect of cleanliness and a sluggish circulation; the patients are generally anæmic, and often scrofulous, and in such children the condition is often the result of some more general ophthalmia.

In adults severe sycosis of the eyelids may accompany sycosis of the beard, but, as a rule, no special tendency to disease of the skin is observed.

Treatment.—When the inflammatory symptoms are severe nothing has such a marked effect as pulling out all the lashes. Cases of a few weeks' standing may be cured by one or two such epilations, together with local remedies, and in old cases it gives great relief in the relapses which are so common. Local applications are always needed (1) for the removal of the scabs, (2) to subdue the inflammatory symptoms. The first object is gained by the use of a weak alkaline lotion, with which the lids are to be carefully soaked for a quarter of an hour night and morning; it is best to use it warm, and the addition of a little tar solution adds to its usefulness. A weak mercurial ointment is to be applied along the edges of the lid after each bathing. This plan of treatment is efficient if the mother will take pains. Another plan, which may be combined with the above, consists in painting the border of the lid with nitrate of silver, in strong solution or in the use of weak silver lotion; but it is not necessary except in severe cases. In old cases with much epiphora the canaliculus is to be slit up. The patients generally need a long course of iron.

A stye is the result of acute suppurative inflammation of one of the glands in the margin of the lid. Owing to the close texture of the tarsus and the vascularity of the parts the pain and swelling are often disproportionately severe; the patient is sometimes frightened by the great swelling of the whole lid which precedes the suppuration. The matter generally points along the border of the lid—sometimes actually around one of the eyelashes; but if a meibomian gland deep in the lid is involved, it will point on the conjunctival or sometimes on the skin surface.

Styes almost always show some derangement of
health. Over-use of the eyes, especially if hypermetropic, is the exciting cause in some cases;
exposure to cold wind in others. Styes are very apt
to occur one after another in successive crops for
several weeks.

Treatment.—A stye may sometimes be cut short if
seen quite early by the vigorous use of an antiphlogistic lotion. A little later the attack may be
shortened by thrusting a fine point of nitrate of
silver into the orifice of the gland if this can be
identified, the corresponding eyelash being first
drawn out. But often poulticing gives most relief
until the stye points, when it should be opened. The
health always needs attending to, and a purgative
iron mixture often suits better than anything else.

Some persons are subject to very small pustules or
styes much more superficial than the above, and less
closely associated with derangement of health.

A meibomian gland is often the seat of chronic
overgrowth, a little tumour in the substance of the
lid being the result (*meibomian cyst, chalazion*). In
a few weeks or months the growth becomes as large
as a pea or bean, and is noticed as a hemispherical
swelling beneath the skin. It generally causes
thinning of the tissues towards the conjunctiva, and
is recognised by a bluish patch on the inner surface
of the lid. As the deeper part of the gland is
generally affected, there is seldom any swelling of
the border of the lid; and when seated close to the
border, as sometimes happens, the tumour is usually
smaller. The skin is freely moveable over the
tumour, but occasionally the growth pushes forwards
and adhesion occurs; but even then it is easily distinguished from a sebaceous cyst by the firmness of
its deep attachment. During its course the meibomian tumour often inflames and sometimes sup-
purates, and in such a case it will become a " stye "

if near the border. The same tumour may inflame several times, and finally suppurate and shrink. Like styes, these tumours are apt to continue forming one after another. They are much commoner in young adults than earlier or later; they are now and then seen in infants. Patients as often apply for the disfigurement as for any discomfort which these little growths occasion.

Treatment.—The cyst is to be removed from the *inner surface* of the lid; in the rare cases where it points forwards the incision must be in the skin; they never recur. The tumour generally consists of a very soft pinkish, somewhat gelatinous, mass or of a gruelly fluid, or may be puriform (*see* Operations).

Small yellow dots are sometimes seen on the inner surface of the lids due to little cheesy collections in the follicles of meibomian glands, and causing irritation by their hardness. They should be picked out with the point of a knife.

Warty formations are not very common on the border of the lid, and are of little consequence excepting in elderly people, when they should be looked upon with suspicion as the possible starting-points of rodent cancer. A small fleshy, yellowish-red flattened growth is sometimes met with just upon the tarsal border causing some irritation; it should be pared off.

Small cutaneous horns are occasionally seen on the skin of the eyelids; they should be snipped off.

Molluscum contagiosum comes under the notice of the ophthalmic surgeon, because the eyelids are so often its seat. One or more little rounded prominences are seen in the skin, varying from the size of a mustard seed to a cherry, but usually not larger than a sweet pea; at first hemispherical, afterwards becoming constricted at the base, and always showing a small dimpled orifice at the top, which may be plugged by dried sebaceous matter. The skin is tightly stretched, thinned, and adherent. The larger specimens sometimes inflame, and may then, without due care, be mistaken. Each molluscum must be removed, the white lobulated, enlarged sebaceous gland which forms the growth being squeezed out through the incision made by a knife or scissors.

The form of acne of the eyelids called *milium* scarcely needs mention here, since patients do not come to us for it. The same is almost as true of *xanthelasma*, which appears as one or more yellow patches like pieces of washleather in the skin, varying from mere dots to the size of a kidney bean, quite soft in texture, and only a very little raised. They are commonest near the inner canthus, and unless symmetrical are usually on the left side. They are never seen in young people; their subjects have usually been liable to become often very dark around the eyes when out of health. The patches are due to infiltration of the deeper parts of the skin by groups of cells loaded with yellow fat. The frequency of xanthelasma in the eyelids is, perhaps, related to the normal presence of certain peculiar granular cells, some of which contain pigment, in the skin of these parts.

The *pediculus pubis* (crab louse), if it happens to reach the eyelashes, will flourish there. The lice themselves cling close to the border of the lid, and look like little dirty scabs. The eggs are darkish, and may also be mistaken for bits of dirt. The absence of inflammation and the rather peculiar appearances will lead, in cases of doubt, to the use of a magnifying glass, which will settle the question at once.

Ulcers on the eyelids may be malignant or lupous or syphilitic, and in the last case the sore may be either a chancre or a tertiary ulcer.

Rodent cancer (rodent ulcer, flat epithelial cancer) is by far the commonest form of carcinoma affecting the eyelids, although cases are occasionally seen of which both the clinical and pathological characters are those of ordinary epithelioma. The peculiarities of rodent cancer are, that it is very slow, that ulceration almost keeps pace with the new growth, and that it does not cause infection of lymphatics. It seldom begins before, generally not until considerably after, middle life, and its course often extends over many years. Beginning as a "pimple" or "wart," it gradually spreads, but some years may elapse before it is as large as a sixpence. When first seen we generally find a shallow ulcer involving the border of the lid, covered by thin scab, bounded by a raised sinuous edge, which is nodular and very hard, but neither inflamed nor tender. Slowly extending both in area and depth, it attacks all tissues alike, finally destroying the eyeball and opening into the nose. In a few very chronic cases the disease remains quite superficial, and cicatrisation may occur at some parts of the ulcerated surface. Now and then a considerable nodule of growth forms in the skin before ulceration begins.

The diagnosis is generally quite easy. A long-standing ulcer of the eyelids in an adult is nearly certain to be rodent cancer. *Tertiary syphilitic ulcers* are much less chronic, more inflamed and punched out, and devoid of the very peculiar hard edge of rodent ulcer, and they are very uncommon. *Lupus* seldom occurs so late in life as rodent cancer, presents more inflammation and much less hardness, and is often accompanied by lupus elsewhere on the cutaneous or mucous surfaces. Lupus is seldom difficult to distinguish on the eyelids from tertiary syphilis, the latter being more acute, more dusky, and showing more loss of substance, with none of the little ill-defined soft tubercles seen in lupus.

When *chancres* occur on the eyelid the induration

4

and swelling are always very marked, the surface
being abraded and moist, but very little ulcerated ;
the glands in front of the ear and behind the jaw
become much enlarged. The same glands enlarge,
either with or without suppuration, in lupus of the
lid and in many inflammatory conditions.

Treatment of rodent cancer.—Early removal by the
knife is of great importance, and probably the more so
in proportion to the youth of the patient. Chloride-
of-zinc paste or the actual cautery are necessary in
addition to the knife in bad cases. The disease is
very apt to return locally. Even in very advanced
cases, where complete removal is impossible, the
patient may be made much more comfortable, and
life probably prolonged by vigorous and repeated
treatment.

Congenital ptosis is a not very uncommon affection.
It is seldom symmetrical, is stated to have been
present at birth, and its causation is obscure. It
sometimes diminishes markedly in the first few years
of life, and this, taken with the fact that we never
see it in grown-up persons, make it probable that
gradual spontaneous recovery is the rule. It is customary to remove an elliptical piece of skin from the
lid for its alleviation, and the result, though often
disappointing, is not injurious unless a very large
piece be excised.

Epicanthus is a rare condition, in which a fold of
skin between the inner end of the brow and the side
of the nose stretches across and hides the inner canthus. If it does not disappear as the child's nose
developes, an operation (removal of a piece of skin
from the bridge of the nose) is indicated.

CHAPTER V

DISEASES OF THE LACHRYMAL APPARATUS

May be divided into those which affect the secret-
ing apparatus—the lachrymal gland and its ducts ;
and those in which the drainage arrangements
situated at the opposite side of the orbit are at fault
—the puncta, canaliculi, lachrymal sac, and nasal
duct. In the great majority of cases the fault lies
entirely in the drainage system.

(1.) The *lachrymal gland* may be the seat of acute or
of chronic inflammation, and in either case an abscess
may be the result. In chronic cases the enlarged
gland is distinctly felt projecting, and can generally
be recognised by its well-defined and lobulated
border ; but in such cases the enlargement cannot
always be distinguished from that caused by a
morbid growth in the gland or corresponding part
of the orbit. In acute inflammation there are the usual
signs, local heat, tenderness, and pain, with swelling which may obscure the boundaries of the gland.
In all cases, if the enlargement is great, the eyeball
is displaced, and the oculo-palpebral fold of conjunctiva in front of the gland is pushed downwards, so
as to project more or less between the lid and the eye.
When an abscess forms it may sometimes be opened
from the conjunctiva, but more often it points to the
skin, through which the incision must then be made
(*see* Operations). If the abscess be allowed to
break spontaneously through the skin a troublesome
fistula may be the result.

A little abscess sometimes forms in one of the separate lobules, forming the anterior part of the gland, the main body of the gland (the "superior lachrymal gland") remaining free.   There is limited swelling and tenderness of the deep lid structures at the upper outer angle, not passing back beneath the orbital rim ; the yellow head of the abscess may be seen pointing through the conjunctiva, near the oculo-palpebral fold above the tarsal cartilage, a position which serves to distinguish this abscess from a suppurating meibomian cyst.   Inflammation of the lachrymal gland, whether acute or chronic, is commoner in children than adults, and, I believe, in women than in men.

Cystic distension of one or more of the gland ducts is now and then seen in the form of a bluish semi-transparent swelling (Dacryops), just beneath the conjunctiva of the lid at the upper-outer part.   Functional diseases of the gland are rare ; a few cases of epiphora, in which the symptom is subject to variations, and where there is no obstruction in the drainage system, are, perhaps, instances of hyper-secretion.   No change appears to have been noticed in cases of paralysis of the cervical sympathetic nerve.

The flow of tears over the edge of the lid on to the cheek has been called *epiphora* when it is the result of over-action in the gland, the term *stillicidium lachrymarum* being used when the same symptom is caused by obstruction to the outflow.   No useful purpose is served by keeping the two names, and only the former will be used.   *Lachrymation* is a convenient term for the increased flow which accompanies superficial inflammation of the eyeball.

(2.) *The drainage system* may be at fault in any part from the puncta to the lower end of the nasal duct.

The slightest change in the position of the lower

punctum causes epiphora. In health the punctum is
directed backwards against the eye; if it looks
upwards or forwards the tears do not all reach it,
and some of the secretion will then flow over a lower
part of the lid. In paralysis of the facial nerve the
patient sometimes comes to us for epiphora; the
symptom is caused partly by loss of the compressing and sucking action effected by winking, partly
by a slight falling of the lid away from the eye,
and a consequent change in the position of the
punctum. These patients sometimes notice the
"watery eye" before they discover the other symptoms. The various chronic diseases of the border
of the lids (ophthalmia tarsi), and also granular
disease of the conjunctiva (granular lids), are fer-
tile sources of (1) tumefaction with narrowing, (2)
cicatricial stricture, of the puncta and of the canaliculi; and in both cases the puncta are displaced
as well as constricted. Narrowing, even to complete
obliteration, of the puncta, may occur in children as
the result of former inflammation, of which all traces
have long since passed away. Wounds by which
the canaliculi are cut across generally cause their
obliteration, epiphora being the result.

In all the above cases the epiphora is accompanied
by a visible change in the size or position of the
punctum, none of the symptoms of inflammation
or stricture in the lachrymal sac or nasal duct being
present. Simple division of the canaliculus will
cure, or at least relieve, the watering of all these
cases (*see* Operations), but it is seldom necessary in
the epiphora of facial paralysis.

Epiphora not explained by any of the above
changes is in most cases caused by stricture of the
nasal duct, with or without disease of the lachrymal sac.

Primary disease of the sac, excepting as the result
of a morbid growth, is not common; most diseases
of the sac are due either to retention of secre-

tion by stricture of the duct below, or, more rarely, to participation of its mucous membrane in some chronic inflammation, which affects the conjunctiva or the Schneiderian membrane.

Stricture of the nasal duct is generally caused by chronic thickening of the mucous and submucous tissues lining the canal, but in a large number of cases no cause can be assigned why this change should take place ; it is commonest after middle life. In some cases the change evidently forms a part of chronic disease of the neighbouring mucous membranes. In another group the stricture is the result of periostitis or of necrosis, and of these changes syphilis (either acquired or inherited), scarlet fever, and smallpox are the commonest causes. Injuries to the nose account for a certain number of cases.

The stricture may be seated at any part of the duct, but the upper end, where there is often a natural narrowing, is the commonest spot. It may be fibrous and firm, or soft and due to thickening of the mucous membrane alone.

Stricture of the nasal duct by preventing the escape of tears from the lachrymal sac leads to its distension, to chronic thickening of its lining membrane, and increased secretion of mucus. The mucus may be clear or turbid. A point is reached at length when the distension can be seen as a little swelling under the skin at the inner canthus (*mucocele* or *chronic dacryo-cystitis*). This swelling can generally be dispersed by pressure with the finger, the mucus and tears either regurgitating through the canaliculus or being forced through the stricture down into the nose. In cases of old standing the mucous membrane is often much thickened, and the swelling cannot then be entirely dispersed by pressure.

A mucocele is always very apt to inflame and suppurate, the result being a *lachrymal abscess*. Most cases of lachrymal abscess, indeed, have been preceded by mucocele. The formation of this abscess

gives rise to great pain and tense, brawny, dusky
swelling, which extends for a considerable distance
around the sac, and is sometimes mistaken for erysipelas. The matter always points a little below the
tendo palpebrarum ; for this reason, and because the
pus often burrows in front of the sac, forming little
pouches in the cellular tissue, the abscess should be
opened through the skin, not by slitting the canaliculus.* The incision is to be nearly vertical, but
curving a little outwards. It is always well to
examine for bare bone at the same time. If the
abscess be allowed to break a fistula is very likely to
form. When the parts have quieted down the former
condition of mucocele will be reproduced, and another
abscess may form at any time unless free permanent
exit be given for the secretion.

Treatment.—The object aimed at is the permanent
dilatation of the stricture, but whether this can be
gained or not a free opening should always be made
into the sac from the canaliculus, that retained
secretion may be often and easily squeezed out. The
diligent and long use of astringent lotions to the conjunctiva is also useful, particularly in soft strictures,
some of the lotion reaching the sac and duct.

Dilatation by probing (*see* Operations) is the ordinary and best treatment for all strictures, whether
there be mucocele or not, the rule being to use the
largest probe that will pass readily. The probing is
repeated every few days or less often according to
the duration of its effect, and often needs to be continued for weeks or months. When, as is common,
the stricture is tough, and its dilatation difficult,
it often contracts again ; and it is then customary
to divide the stricture by thrusting a strong-backed,
narrow knife down the duct. It must be confessed,
however, that in a considerable proportion of cases,
whether the stricture be soft or firm, the results of
all treatment are very discouraging, and that the

* But some surgeons prefer the latter position.

benefit obtained is not always worth the pain and inconvenience. In cases where the stricture is quite soft and the obstruction due rather to general thickening of the mucous membrane and over-secretion of mucus than to submucous thickening, the occasional passage of a very large probe is useful, or washing out the duct with astringents by means of a specially constructed syringe is sometimes practised. In cases of long standing, where all other treatment has failed and the lachrymal sac is much thickened, its complete obliteration by the actual cautery gives great relief, or extirpation of the lachrymal gland may be practised. For refractory children and for patients who cannot be seen often, a style of silver or lead, passed in exactly the same way as a probe, but worn constantly for many weeks, is sometimes very useful; but it may slip into the sac out of reach unless furnished with a bend or head so large as to be somewhat unsightly. Neither probes nor styles should be used until the inflammatory thickening and tenderness following a lachrymal abscess have subsided.

Suppuration of the lachrymal sac, on one or both sides, sometimes takes place in new-born infants without apparent cause ; if there is much redness the abscess should be opened, but the suppuration is sometimes chronic and will cease under the use of astringent lotions. Some cases of epiphora, with contracted punctum in older children, may be the consequences of this infantile suppuration. Cases in which the sac or duct is obliterated by injury can seldom be relieved.

CHAPTER VI

DISEASES OF THE CONJUNCTIVA

May be divided into those which from the outset are general and affect the whole membrane, ocular and palpebral alike, and of which the various forms of contagious ophthalmia are examples ; and those which primarily affect either the ocular or the palpebral part alone. *The term "ophthalmia" includes all inflammations of the conjunctiva, and should not be applied to any other diseases.*

GENERAL DISEASES

The conjunctiva, like the urethra, is subject to purulent inflammation, and, like the respiratory mucous membrane, is liable to the muco-purulent, and to the membranous or diphtheritic forms of disease. All cases in which there is yellow discharge are in greater or less degree contagious. The congestion, which forms a part of conjunctivitis, is much influenced by age ; the younger the patient the less is the congestion in proportion to the discharge, a fact which is to be borne in mind in examining patients at both ends of the scale.

Purulent ophthalmia (O. neo-natorum, Gonorrhœal O., Blenorrhœa of the conjunctiva) is generally due to contagion from the same disease, or from an acute or chronic discharge from the urethra or vagina, whether gonorrhœal or not. Muco-purulent ophthalmia when quickly passed on from one to another under conditions of health favorable to suppuration (*e. g.* weakness after acute exanthems) may be intensified into the purulent form. Gonorrhœa has been experimentally produced by inoculation with pus from purulent ophthalmia. Some animals ·are subject to purulent ophthalmia, but it is said that the discharge from the human disease, and even from gonorrhœa, gives no result on the conjunctiva of rabbits. Like gonorrhœa, purulent ophthalmia may occur more than once, and varies greatly in severity in different persons. The quality of the infecting discharge no doubt has much influence, severe forms being generally caused by discharge from a recent or severe source of infection; but chronic discharge may give rise to a severe attack. The health of the recipient and the previous condition of the eyelids exert an important influence; if the lids are granular, slight causes—for example, trivial blows—sometimes bring on severe purulent ophthalmia.

The disease sets in from twelve to about forty-eight hours after inoculation; in infants the third day after birth is almost invariably given as the date when discharge was first noticed. Itchiness and slight redness of conjunctiva soon pass on to intense congestion with chemosis of conjunctiva, tense inflammatory swelling of the lids, great pain, and discharge. The discharge at first is serous, or like turbid whey, but soon becomes more profuse and uniformly creamy (purulent) and yellow, or even slightly greenish. Dark abrupt ecchymoses are often present.

The lids are always swollen, hot, and red, and in bad cases very tense and dusky. The upper lid hangs down over the lower, and is often so stiff that it cannot be completely everted. The conjunctiva is succulent and easily bleeds. The disease if untreated declines spontaneously, and the discharge almost ceases in about six weeks, the palpebral conjunctiva being left thick, relaxed, and more or less granular.

Cicatricial changes, identical with, but less severe than, those resulting from chronic granular lids, and analogous to what occurs in stricture of the urethra, sometimes follow; in others a leathery thickening of the ocular conjunctiva is noticed.

There is a great risk to the cornea in this disease, chiefly from strangulation of the vessels, but also from the influence of the discharge. If within the first two or three days the cornea becomes hazy and dull, like the eye of a dead fish, there is great risk that total or extensive sloughing will occur. In milder cases almost transparent ulcers may form near the margin, and, quickly eating into the cornea, sometimes cause perforation. In many of the slighter cases, such as are seen in infants, no corneal damage occurs. Either one or both eyes may be attacked; in adults one eye often escapes; in infants, where the inoculation occurs during birth, both eyes almost always suffer.

Treatment.—If only one eye is affected, and the patient old enough to obey orders, the sound eye must be sealed up with pad and bandage covered by collodion, or, much better, with the shield introduced by Dr. Buller, which consists of a piece of macintosh about 4½ in. square, with a watch glass fastened into a hole in the centre, through which the patient can see; this is fixed by broad pieces of strapping to the nose, forehead, and cheek, its lower-outer angle being left open for ventilation; particular attention is to be paid to the fastening on the nose. All concerned are to be warned as to the risk of contagion and the means of conveyance. The essential curative measures are—(1) frequent removal of the discharge by the free use of water. Every hour, day and night, or in adults every two hours, the discharge is removed with soft bits of moistened rag or cotton wool, the lids being gently opened, and a stream of water allowed to trickle freely in; or a syringe or irrigation apparatus may be used. In adults, where the swelling is often extreme and very brawny, we may increase the congestion and irritability by interfering oftener than every two hours. (2) The frequent anointing of the lids with a simple ointment. (3) The use of astringent or caustic lotions or drops to stay the formation of pus, and to be used from every hour to every two or three hours, according to the case and the nature and strength of the solution. The lotion may be alum (eight or ten grains to the ounce), or sulphate of zinc and alum, used very freely every hour or two, or corrosive sublimate (one eighth or one quarter of a grain), or chloride of zinc (two grains, with just enough dilute hydrochloric acid to make a clear solution), used freely every two or three hours, or nitrate of silver (two grains), four or six times a day. Many surgeons greatly prefer the last to all others. (4) Strong solutions of nitrate of silver, or the mitigated solid nitrate, are of great service in shortening the attack and lessening the risks, and should be used in all severe cases unless specially contraindicated. A ten- or twenty-grain solution is brushed freely over the conjunctiva of the lids everted as well as possible, and freed from discharge. If the mitigated stick is used more care is needed; and to prevent too great an effect, it is to be washed off with water after waiting about fifteen seconds. These strong applications must be made by the surgeon. The pain caused by them is lessened and the beneficial effect increased by free bathing with cold or iced water afterwards. The application is not to be repeated until the discharge, which will be markedly lessened for some hours, has begun to increase again, and is seldom needful or justifiable more than once a day. (5) Local cold by iced water or thin iced compresses; in severe cases to be used almost constantly, in milder cases for frequent periods of half an hour. This plan, but little adopted in our hospital practice, is very highly spoken of as most efficacious, if efficiently carried out, but if only half done it is to say the least useless. (6) In the early stage, in adults, several leeches to the temple will give relief, or, if the swelling be very tense, we may divide the outer canthus with scissors or knife, and thus both bleed and relax the parts at the same time. Scarification of the palpebral conjunctiva and radial incisions in the ocular conjunctiva are sometimes very useful. (7) Hot fomentations are sometimes better than cold, the heat perhaps acting as a counter-irritant to the conjunctiva by increasing the congestion of the skin and simultaneously relieving its tension.

The following additional precautions are important :—Strong nitrate of silver applications are unsafe in the earliest stage, before free discharge has set in, and also in cases where, even later in the disease, there is much hard brawny swelling of the ocular conjunctiva and comparatively little discharge ; cases, in fact, approaching the condition known as diphtheritic ophthalmia. In these either very cold or very hot applications, leeches, cleanliness and weak lotions, should be chiefly relied upon. Purulent ophthalmia is generally worse in adults than in infants. The ice and leeching treatment are seldom advisable for infants. It is of extreme importance to begin treatment very early, for the cornea is often irreparably damaged within two or three days. The patients, if adults, are often in feeble health, and need supporting treatment. Ulceration of the cornea does not contraindicate the use of strong nitrate of silver if the discharge is abundant. Treatment must be continued so long as there is any discharge or the conjunctiva of the lids remains fleshy, for a relapse of purulent discharge often takes place if remedies are discontinued too soon.

Muco-purulent ophthalmia.—The commonest and best characterised of the acute ophthalmiæ is the so-called *catarrhal ophthalmia.* The name is a bad one, for neither does the disease form part of a general catarrh of the respiratory tract, nor does it show the tendency to relapse so characteristie of catarrh, nor does it seem to be caused by cold. This ophthalmia attains its height very quickly, almost always attacks both eyes, and spontaneous recovery takes place in about a fortnight. There is great congestion, much gritty pain, which often prevents sleep, spasm of the lids, free muco-purulent discharge, and, in many cases, ecchymotic or thrombotic patches in the conjunctiva. The lids are somewhat swollen and red, but never tense, and the cornea seldom suffers.

This disease is very contagious, far more so than purulent ophthalmia, for which it is now and then mistaken.

It varies much in severity, even in different members of the same household who catch it almost at the same time, but attacks all ages indifferently. It is, I believe, commonest in warm weather, or perhaps at the change from cold to warm. It is rare to find that the patient has suffered from the disease before. Any mild astringent lotion will cut it short.

Troublesome *ophthalmia, with muco-purulent discharge*, is common in children *after exanthemata*, especially measles. It runs a far less definite course than the preceding disease, shows but little tendency to spontaneous cure, and is very often complicated with phlyctenular ulcers of the cornea and sycosis and excoriations of the eyelids ; the patients are frequently strumous. The discharge is seldom so abundant as in the disease just considered. The treatment is often troublesome, and many changes have to be tried ; weak astringents or weak alkaline lotions of soda or borax, with the use of the yellow ointment, both to the skin and conjunctiva ; sometimes calomel dusted into the eye or weak nitrate of silver lotions will be the best local means ; atropine often increases the irritation. Of course careful attention to health is necessary. The patients should not be confined to the house, but, with a large shade over both eyes, should take plenty of exercise in fine weather. The eyes should not be bandaged.

Some forms of acute conjunctivitis, with little or no discharge, are seen both in children and adults, which do not conform to the above types, and are of comparatively slight importance. Many such appear to depend on changes of weather or exposure to cold, and are complicated with phlyctenulæ. A few are distinctly rheumatic. The conjunctiva is involved more or less in herpes zoster of the ophthalmic division of the fifth nerve in erysipelas of the face, and in the early stage of measles. Slight degrees of chronic conjunctivitis are set up by many continuous local irritants, dust, smoke, cold wind, &c., and by the strain attending the use of the eyes without glasses in cases of hypermetropia. Mention must be made of the not very common cases in children, where an ophthalmia appears to form part of an impetiginous or herpetic eruption on the face, with which it is simultaneous. These differ from the ordinary instances in which the lids, cheek, and lining membrane of the nose are irritated into an eruption by tears and discharge from a pre-existing conjunctivitis.

Muco-purulent ophthalmia, of any kind, becomes a very important affair if it breaks out in schools or armies, &c., where granular disease of the eyelids is prevalent (see p. 66).

Membranous and diphtheritic ophthalmia. In a few cases of ophthalmia, either purulent or muco-purulent, the discharge adheres to the conjunctiva in the form of a membrane (*membranous or croupous ophthalmia*). Still more rarely, in addition to adherent membrane on the surface, the whole depth of the thickened conjunctiva is stiffened by solid exudation, which much impairs the mobility both of the

lids and eyeball, and by compressing the vessels prevents the formation of free discharge, and places the
nutrition of the cornea in great peril. It is to the
latter cases that the term " diphtheritic " is limited by
most authors; but we find many connecting links
between the two types above defined, and between
each of them and the ordinary purulent and muco-
pûrulent cases.

It is of much consequence in practice, both for
prognosis and treatment, to recognise the presence of
membranous discharge and of solid infiltration in
any case of ophthalmia ; the liability to severe corneal damage being much increased by both these
conditions, and especially by the latter.   When membrane is present, it may cover the whole inside of the
lids, or it may occur in separate or in confluent patches; it often begins at the border of the lid, and
is seldom found on the ocular conjunctiva.   It can
be peeled off, and the conjunctiva beneath bleeds
freely, unless infiltrated and solid ; in the latter case
the membrane is more adherent, and the conjunctiva
is of a palish colour, and scarcely bleeds when exposed.   The firmer the membrane, or the greater the
solid infiltration, the less is there of purulent discharge, though the natural tendency in most cases is
for the solid products, whether membrane or deep
infiltration, to pass after some days into a stage of
liquefaction, with free purulent secretion.   In rare
cases the membrane forms and re-forms for months.
As regards cause, (1) very rarely the process creeps
up to the conjunctiva from the nose in cases of
primary diphtheria, or it may be caused by inoculation of the conjunctiva with membrane; whilst
in a few the ophthalmia forms the first symptom of
general diphtheria or of masked or anomalous
scarlet fever.   (2) Much more commonly it is
part of a diphtheritic type of inflammation following some acute illness.   (3) It may be caused by
the over-use of caustics in ordinary purulent oph-

thalmia (p. 61). (4) It may be due to contagion, either from a similar case or from a purulent ophthalmia, or a gonorrhœa, the membranous or diphtheritic type depending on some peculiarity in the health or tissues of the recipient. Membranous and diphtheritic ophthalmia are seen most often in children from two to eight years old, sometimes in young infants, and less commonly in adults. The worst cases are much commoner in North Germany than in other parts of Europe.

In *treatment* the cardinal point is not to use nitrate of silver in any form when there is much solid infiltration of the conjunctiva and but scanty discharge. The agents to be relied upon are (1) either ice or hot fomentations ; ice, if it can be used continuously and well ; fomentations, to encourage liquid exudations and determination to the skin if the cold treatment cannot be carried out, or fails to make any impression on the case ; (2) leeches, if the patient's state will bear them ; (3) great cleanliness. The presence of membrane is no bar to the use of caustics, provided that the conjunctiva is succulent, red, and bleeds easily.

The use of atropine sometimes gives rise to a peculiar irritation and inflammation of the conjunctiva and skin of the lids—*atropine irritation.* The conjunctiva is reddened, and on the lids it becomes thickened, and even granular. The skin is reddened, somewhat shining, though lax, and whilst not losing its wrinkles, it becomes glazed and slightly excoriated. This effect of atropine is commonest in old people. Some persons are very susceptible and cannot bear even a drop or two without suffering in some degree. Daturine is to be used instead of atropine, unless it be safe to disuse all mydriatics for a few days. An ointment containing some lead and zinc should be applied to the lids, and an astringent zinc lotion to the conjunctiva; in other cases glycerine to the skin is better than anything, and sometimes a bread poultice gives most relief.

Granular ophthalmia (trachoma) is a very important malady, characterised by slowly progressive changes in the conjunctiva of the eyelids, during which it becomes thickened, vascular, and roughened by firm elevations, instead of being pale, thin, and smooth. The change usually begins in the follicular structures of the conjunctiva of the lower lid, extending to the papillæ and the submucous tissue of both lids at a later period, and giving rise to the growth of much organised new tissue in the deep parts of the conjunctiva. This tissue is afterwards partly absorbed and partly converted into dense tendinous scar tissue, which by very slow shrinking often gives rise to much trouble. It is important to remember that the conjunctiva in this disease does not ulcerate, and that the prominences are not " granulations " in the pathological sense.

The disease is at first shown by the presence, on the lower lid, of a number of rounded, pale, semitransparent bodies like little grains of boiled sago, or sometimes looking like vesicles; these are the so-called " vesicular," or sago-grain, or " follicular " granulations (Fig. 10). Many, if not all of them are

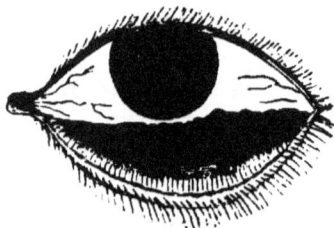

FIG. 10.—Granular lower lid (after Eble).

lymphatic follicles. They are, to a certain but varying degree, normal structures, and are seen in greater or less numbers, especially on the lower lids, in many young persons with slight ophthalmia who never afterwards suffer from true granular lids. Such mild cases in which no parts deeper than the follicles and papillæ are affected, and in which recovery takes place without cicatricial changes, are by some distinguished authors placed, under the name of *conjunctivitis follicularis*, in a separate category from the granular disease. The latter disease is held on this hypothesis to depend on a different morbid process, the growths or " granulations " bearing no relation to lymph follicles. But the frequent coincidence of both the changes in the same case, the fact that both admittedly occur under the same general conditions, and that in a given case the distinctions between " follicles " and " granulations " often cannot be made until it is known whether or not cicatricial changes will occur, certainly much lessen the clinical value of what may, or may not, turn out to be a real pathological difference.

Granular disease is important chiefly because it greatly increases the susceptibility of the conjunctiva to take on acute inflammation and to produce contagious discharge, makes it less amenable to treatment, and very liable to relapses of ophthalmia for many years. Indeed, so vulnerable is the conjunctiva in this state that it is rare in ordinary practice to see granular lids of long standing without the history of an acute ophthalmia at some previous time, though many such may be seen in crowded schools, &c.

Chronic granular disease is the result (1) of prolonged overcrowding, or rather of long residence in badly ventilated and damp rooms, and hence it is seen in the greatest perfection in workhouse schools, reformatories, jails, and barracks ; (2) a generally low state of health, no doubt, increases the susceptibility to it ; (3) it is, *cœteris paribus*, commonest and most quickly produced in children ; (4) certain races are peculiarly likely to suffer, *e. g.* the Irish, the Jews, and some other Eastern races, and some of the German and French races. The Irish and Jews carry it with them all over the world, and transmit the liability to their descendants wherever they live. Negroes in America are said to be almost exempt ; (5) damp and low-lying climates are more productive of it than others, and possibly what are now race tendencies may be the expression of climatal conditions acting on the same race through many generations. When accompanied by discharge the disease is contagious, but not otherwise. It is generally held that the discharge from a case of trachoma is specific, *i. e.* that it will give rise by contagion, not only to acute ophthalmia, but to the true granular disease. Owing to numerous fallacies this point is a very difficult one to decide.

Those who practise in the army, especially in India or Malta, or who have charge of such institutions as pauper schools, will find that in practice the causes of the chronic granular state are inextricably mixed up with all kinds of facilities for contagion, and that it will be necessary to fight against two enemies—the causes of spontaneous granular disease, and the sources of contagious discharge. The former is to be combated by improved hygienic conditions, especially by free ventilation, dry air, abundant open-air exercise, and improving the general vigour. The sources of contagion are endless, especially since, as has been stated, granular patients are liable to many relapses of muco-purulent discharge from almost any slight irritation. Frequent inspection of all the eyes, and rigid separation of all who show any discharge or are known as especially subject to relapses ; such arrangements for washing as will prevent the use in common of the towels and water ; and extreme care against the introduction of contagious cases from without, are the chief preventive measures. Extra precautions will be needed in time of war or

famine, or when measles or scarlet fever are prevalent,
or while marching in hot, sandy, or windy districts.

The *curative treatment* when discharge is present
does not differ from that of the acute ophthalmiæ
already given.   The use of strong astringents (solid
sulphate of copper) or caustics (nitrate of silver in
strong solutions, or in the mitigated solid pencil),
however, is generally needed in order to make
much impression on the granular state of the lids.
The lids, being thoroughly everted, are touched
with one or other application, and this is repeated
daily, or three times a week or less often, according
to experience.   Some practice is required before we
can decide on the needful frequency for each case.
By careful treatment on this principle most patients
may be kept comfortably free from active symptoms,
many relapses prevented, the duration of the disease
shortened, and the risks of secondary damage to the
cornea greatly lessened.   Do what we will, however,
granular disease when well formed is most tedious,
and fastens many risks and disabilities on the patient for years to come.

For routine treatment on a large scale nothing is
so effectual as nitrate of silver, either a ten- or twenty-
grain solution, or the mitigated solid point.   But silver
has the disadvantage of sometimes permanently staining the conjunctiva after long use, and in very chronic
cases I think either sulphate of copper or lapis
divinus is to be preferred, especially as the patient
may often be taught to evert his own lids and use it
himself.   The solid mitigated nitrate of silver needs
washing off with water at first (p. 60), but in old
cases it is often better not to do so.

Results of granular disease.—The friction of the granulations on the upper lid (*a*, Fig. 11), especially in cases
of long standing where some scarring is present (*b*),
often causes haziness of the cornea partly from ulceration, but mainly by the growth of a layer of new
and very vascular tissue just beneath the epithelium

(*pannus*) (Fig. 12).   In later periods the conjunctiva
and deeper tissues are shortened and puckered by

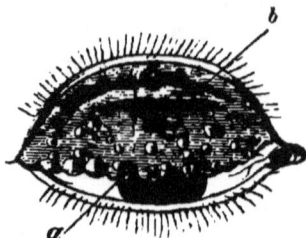

FIG. 11.—Granular upper lid, with scarring.

the scar following absorption of the "granulations;"
this leads to distortion of the border of the lid and
to misdirection of the eyelashes, so that some or all
of them turn more or less inwards and rub against
the cornea—*distichiasis, trichiasis, entropion;* and
these are often combined with pannus.   Pannus
begins beneath the upper lid, its vessels are superficial and continuous with those of the conjunctiva,
and are distributed in relation to the parts covered
by the lid, not in reference to the structure of the
cornea (Fig. 13).   The proper corneal tissue suffers
but little except where ulceration occurs, but when
the vascularity is extreme the cornea may soften and
bulge even without ulcerating.

Pannus disappears when the granular lid or the
displacement of lashes is cured.   Very severe and
universal pannus is sometimes best treated by artificial inoculation with purulent ophthalmia, the inflammation being followed by obliteration of vessels
and clearing of the cornea ; but this treatment needs
great judgment and caution.   Removal of a zone of
conjunctiva and subconjunctival tissue (syndectomy,
peritomy) from around the cornea is free from risk
and sometimes very beneficial in old cases which,
though severe, are not bad enough for inoculation.   In
old cases of granular disease, even where no complications have arisen, the upper lids often droop from

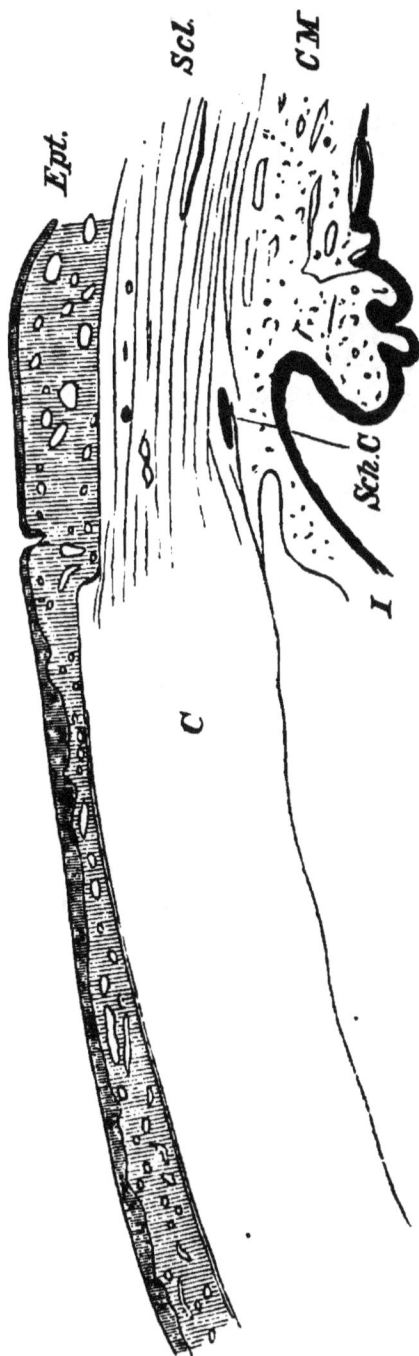

FIG. 12.—Section showing layer of new and vascular tissue (*pannus*) between epithelium (*Ept.*) and cornea (*C.*). *Scl.* sclerotic; *C M.* ciliary muscle; *Sch. C.* Schlemm's canal; *I.* iris.

extension and relaxation of the thickened conjunctiva, the patient acquiring a heavy sleepy look.

For the cure of the displaced lashes and incurved
eyelid we may (1) simply advise repeated epilation
with forceps, (2) extirpate all the lashes by cutting
out the follicles with a narrow strip of the marginal
tissues of the lid (*see* Operations), or (3) by various

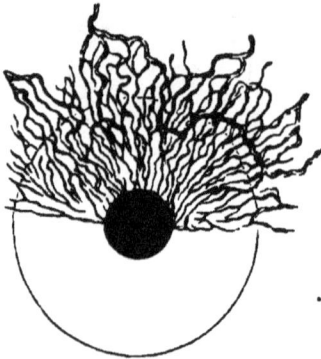

FIG. 13.—Pannus affecting upper half of cornea.

operations we may attempt to restore the parts to
their proper positions.   Many of the numerous operations for restoring the lashes to their normal direction
fail to give permanent relief for various reasons, but
especially because the contraction of the scar continues, and the original state of things is sooner or
later reproduced.

CHAPTER VII

DISEASES OF THE CORNEA

Ulcers and non-specific inflammatory diseases

INFLAMMATION of the cornea may be circumscribed
or diffuse, and may implicate either the proper corneal tissue or the epithelium on one or other of its
surfaces. It may be a local process leading to formation of pus, or to ulceration ; or the expression of a
constitutional disease, such as inherited syphilis ;
or it may form part, and perhaps only a minor part,
of disease involving also the deeper parts of the
eyeball, the iris (kerato-iritis), or ciliary body (cyclo-
keratitis), for example.

The different varieties of corneal ulceration and
suppurative inflammation are among the commonest,
and practically most important, of eye diseases. The
fact that the cornea, although a fibrous structure, is
further removed from the blood-vessels than almost
any other tissue, places its nutrition in particular
danger in all diseased states when the circulation is
feeble or the nutritive quality of the blood below par ;
lastly, the surface of the cornea is so delicate, and
the importance of its perfect transparency so great,
that slight injuries and external irritations are of
more moment·here than in other parts of the body.

When inflamed the cornea always loses its transparency. If only the anterior epithelium is involved,
the surface loses its polish and looks like clear glass
which has been breathed upon—" steamy," or finely
pitted, and this loss of polish is seen in many states of
disease.

Thickening of the epithelium, and still more, exudation into the corneal tissue, gives a white or greyish or yellowish tint to the part. If the corneal tissue is opalescent, while the surface is at the same time "steamy," the term "ground glass" gives a good idea of the appearance, though to make the simile correct the glass ought to be milky throughout as well as ground on the surface. Rapid suppurative inflammation is preceded by a stage of diffused opalescence, and the latter change is therefore a very dangerous sign in such diseases as purulent ophthalmia, or severe burns, or paralysis of the fifth nerve.

Before describing the most important types of corneal ulcer, it is convenient to mention the principal symptoms and appearances attendant on ulceration of the cornea in general. An ulcer of the cornea is preceded by a stage of infiltration, and the inflamed spot is generally a little raised. After the centre of the spot has broken down into an ulcer, some infiltration remains at its base and edges, the quantity and colour of which help us to judge of the probable course of the disease. When the ulcer heals it leaves a hazy or opaque spot (*leucoma* if dense, *nebula* if faint), which is slight and often disappears entirely if superficial, but will be in part permanent if resulting from a deep ulcer. These opacities are likely to clear, *cæteris paribus*, in proportion to the youth of the patient. Time also is a very important element, nebulæ often continuing to clear slowly for years. Lastly, local stimulation aids in the removal of the opacities, one of the best applications being the ointment of yellow oxide of mercury. Some ulcers have scarcely any infiltration, and these for the most part heal slowly and with permanent loss of substance. Such permanent loss of substance, whether with or without opacity, is shown by the presence of a facet or flattened spot on the cornea at the seat of the former ulcer. Such facets, even though quite clear, will, if they are placed opposite to the pupil, interfere with sight more than a nebula which occupies the same position, but is unattended by alteration of the curvature of the cornea.   It is obvious that corneal opacities must be most serious when placed over the pupil.

The chief *symptoms* of corneal ulcerations are (1) *Photophobia*, or at least spasm of the orbicularis, blepharospasm (for it is not always clear whether the centripetal irritation which leads to the closure of the lids, starts from the retina or from the branches of the fifth nerve in the cornea and conjunctiva) ; (2) *Congestion ;* (3) *Pain*.   All three symptoms vary extremely in degree in different cases. As a broad rule, subject to many exceptions, we may say that intolerance of light is stronger in children than in adults, greater in cases of superficial than of deep ulceration, more intense in persons who are strumous and irritable than in those whose tissues are healthy and tone good.   Photophobia should always lead to a careful inspection of the cornea, and we shall then sometimes be surprised to find how slight a change gives rise to this symptom in its severest form.   The degree of congestion differs according to the seat and cause of the ulcer and with the patient's age, being usually greatest in adults. The vessels implicated are very much the same, so far as external examination goes, as in iritis, viz., the small branches beneath the conjunctiva in the ciliary region, the *ciliary zone* (p. 13), but in many cases there is congestion of the conjunctival vessels as well.   In some forms of marginal ulcer, only the vessels which feed the diseased part show any material change.   Great pain often attends the earlier stages of corneal abscess and is common in many acute ulcers; as a symptom, it of course always needs careful attention ; it is generally relieved by the same local measures as are best for the disease itself.

TYPES OF CORNEAL ULCERATION

(1.) One of the simplest forms is the *small central ulcer* often seen in young children. A little greyish-white spot is seen at or near the centre of the cornea, at first elevated and bluntly conical, afterwàrds showing a minute shallow crater ; the congestion and photophobia varying, but often slight. The ulcer is usually single, but is apt to recur in the same or the other eye. The infiltration in many of these cases extends quite into the corneal tissue, and the residual opacity often remains for a long time if not permanently. The patients are almost always badly fed little children. In most cases the ulcer quickly heals, but now and then the infiltration passes into an abscess or a spreading suppurating ulcer.

(2.) In other cases, less common than the above, one or more central ulcers occur and pass through a much more chronic course, and though attended with little infiltration, lead to loss of tissue, so that when the healing process is finished a shallow depression or a flat facet is left with perhaps scarcely any loss of transparency. They occur in older children, and some of the best examples are seen in anæmic or in strumous patients, with granular lids.

(3.) *Phlyctenular ophthalmia* and *phlyctenular ulcers* of cornea (phlyctenulæ, herpes corneæ, pustular ophthalmia, marginal keratitis). The formation of little papules or pustules on or near the corneal margin is exceedingly common, either as an independent event or as a complication of some already existing ophthalmia. Although there are many varieties and degrees of phlyctenular inflammation in respect alike to the seat, the extent, and the course of the disease, the following features are common to all. They show great tendency to recur during several years ; they are seldom seen in very young children, and comparatively seldom after middle life ; they occur so often in strumous subjects that we are justified in strongly suspecting scrofulous tendencies in all who suffer much from them ; ophthalmia tarsi is often seen in the same patients ; the first attack of phlyctenular disease often follows closely after an acute exanthem, and especially after measles ; they are much under the influence of climate.

A little elevated spot, commonly about the size of a small mustard seed, is seen either on the white of the eye near the cornea, or upon, or just within, the corneal border. It is preceded and accompanied by localised congestion. Its apex is sometimes as yellow as the head of an acne pustule, but more commonly it is abraded, flat, and whitish. Pustules at a little distance from the cornea, although generally larger than those seated on the corneal border, occasion less photophobia, and are more easily cured. Pustules at the corneal border, though often very small, commonly give rise to troublesome and sometimes to very severe photophobia; they are troublesome in proportion rather to their number than their size ; if very numerous they run together and form an elevated ring around the cornea, and their cure is then often most tedious.

A pustule is always liable, even when it has begun on the conjunctiva, to extend as a superficial ulcer on to the cornea, though it never extends in the opposite direction over the sclerotic. Such a *phlyctenular ulcer*, if it does not stop near the corneal border, will make, in an almost radial direction, for the centre, carrying with it a leash of vessels which lie upon the track of opacity left in the wake of the ulcer (Fig. 14). Finally, the ulceration stops, the vessels dwindle and disappear, and the path of opacity clears up more or less. The term "recurrent vascular ulcer" is given by some to these ulcers when they are solitary ; they are often, however, multiple as well as recurrent, and the cornea may then finally show a thin and irregular network of superficial vessels on a patchy, uneven, hazy surface, the so-called "*phlyctenular pannus.*"

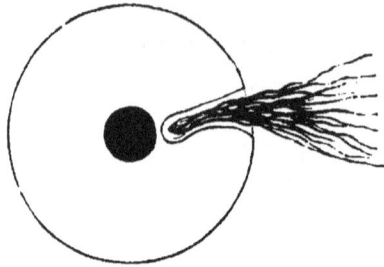

FIG. 14.

A well-defined variety of phlyctenular inflammation, aptly called *marginal keratitis,* is shown in slight degrees in the form of a slight granular-looking swelling all around the edge of the cornea. Vessels soon form, and if the process continues the cornea is encroached on by a densely vascular superficially ulcerated, and yet somewhat thickened zone. In slight degrees this marginal abrasion and thickening is common enough ; severe cases are rare and very serious, leading finally to implication of perhaps the greater part of the cornea. It often begins crescentically above and below, as in Fig. 20.

In another variety a single pustule at the border of the cornea ulcerates deeply, becomes surrounded by swollen and infiltrated tissue, and may perforate ; such cases are seen in weakly women and strumous children.

In very rare cases, what appears to be an ordinary conjunctival pustule persists, grows deeply, and may even perforate the sclerotic in the form of an ulcer ; or it may infiltrate the sclerotic and the ciliary body beneath, forming a soft, semi-suppurating tumour, whence the inflammation is likely to spread to the vitreous and destroy the eye. Stopping short of these extreme results, such a case forms one type of episcleritis.

The corneal changes produced by the friction of granular lids have been considered under that subject. The pannus of granular lids can usually be distinguished from the phlyctenular pannus already mentioned as sometimes following repeated vascular ulcers, by the greater uniformity and closeness of its vessels, and by its being worst under the upper lid (Fig. 13). Any doubt is settled by everting the lid. But it must be borne in mind that ulceration of the cornea often occurs as a complication of trachomatous pannus (p. 70).

(4.) A troublesome and dangerous form of disease, seen for the most part in the senile period of life, is the *serpiginous ulcer*. It is often comparatively chronic. There is much congestion, and often much pain and photophobia. With these symptoms we find either a marginal trough or ditch-like ulcer, with crescentic borders, or a more central ulcer with nearly circular outline, and a varying amount of infiltration of its walls. If the ulcer has lasted some little time one of its borders, the outer, if the ulcer be marginal, will be partly healed and bevelled off, the floor of the ulcer sloping downwards to its inner boundary, which will be sharply cut, or even overhanging.

Slight cases of this kind, taken early, generally give little trouble, especially if the infiltration is but slight. But such an ulcer, if neglected, is very likely to increase in all dimensions, to become complicated with iritis and hypopyon, and to lead to perforation of the cornea; or to spread slowly over the whole cornea, leaving a densely opaque scar in its wake. In either event the eye is much damaged, if not destroyed.

(5.) *Abscess of the cornea* and *acute suppurating ulceration* are common diseases. Abscess may occur at any age, but the greater number of cases are seen in elderly or senile people, and abrasion or some slight injury by a foreign body is not an uncommon cause, especially if near the centre of the cornea. The little grey central ulcers of young children (p. 76) occasionally go on to abscess. It will very often be noticed that the subjects of this form of disease, as well as of the serpiginous ulceration just described, are either senile or under-fed, or if vigorous and full-blooded that they show signs of being damaged by drink. Abscess of the cornea is attended by great pain and congestion, and the case often comes under care pretty early, though often not till the cornea has given way, either in front of or behind the little collection. The spot itself is small, but the whole cornea often becomes more or less hazy, and the purulent infiltration may at a later period spread and involve almost its whole substance.

Hypopyon signifies a collection of pus or puro-lymph at the lowest part of the anterior chamber; its upper boundary is usually but not always level (Fig. 15). It may occur with any rapid form of ulcer,

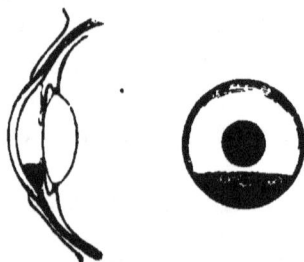

FIG. 15.—Hypopyon, seen from the front, and in section, to show that the pus is behind the cornea.

accompanied by purulent infiltration of the surrounding cornea, whether the ulceration be deep or not, or with corneal abscess, and with any corneal ulcer, chronic or acute, in which iritis supervenes. The pus is derived either from an abscess breaking through the posterior surface of the cornea, or from suppurative action in the layer of epithelium covering Descemet's membrane, or from the surface of the iris, for it is often observed that when the iris inflames in connection with corneal ulceration the iritis is of the same (suppurative) type as the corneal disease. It is to be noted, lastly, that simple iritis now and then gives rise to hypopyon (*see* Rheumatic Iritis).

FIG. 16.—Onyx.

In many cases of severe suppuration (*a*, Fig. 16) the pus sinks down between the lamellæ of the cornea (*b*). To this condition the term *onyx* is applied, and should be limited, though it is sometimes used in other senses; the word is not of much use, and may very well be discarded. It is not always easy to say whether the pus is in the anterior chamber (hypopyon) or in the cornea just in front of the same part (onyx), especially as both conditions are often present together. Hypopyon, when liquid, will change its position if the patient lies down, but as it is more often gelatinous or fibrinous, this test loses much of its value. Oblique illumination will sometimes show the cornea clear in front of an hypopyon. The diameter of the anterior chamber being a little greater than the apparent diameter of the clear cornea, a very small hypopyon is almost hidden behind the overlapping edge of the sclerotic, and may escape detection.

6

Treatment of ulcers of the cornea

The general principles of local treatment applicable in various degrees to the different types of ulceration are :—

(1.) To secure rest by bandaging the eye or shading both eyes, and thus to minimise the movement of the lids over the ulcerated surface; and to soothe local pain by atropine. (2.) To relieve the tension of the ulcerated surface, and so favour healing. Atropine is believed to owe part at least of its good effect in ulcers of the cornea to its asserted power of lessening the tension of the eye. In severer cases this aid is more definitely gained by operative means, which at the same time evacuate an abscess if present (3.) In suppurating cases, to induce granulation instead of suppuration, and absorption of pus already effused. Hot fomentations to the eyelid, often repeated, attain this end better than anything else in a large number of cases. (4.) Stimulation of the surface of the ulcer when it has begun to heal, especially if it is indolent, gives most excellent results. The best stimulants are calomel, yellow oxide of mercury, and nitrate of silver. (5.) Counter-irritation by a seton in the temple is of very great use in some chronic irritable ulcers. (6.) When ulcers are caused by granular lids the treatment of the granular disease is more important than that of the ulceration, unless the latter be of a threatening character.

The choice of one or the other of the above plans of local treatment is easy enough in a large proportion of cases ; in others a good deal of judgment is needed ; while in a certain number it is impossible to say with any certainty what will be found most beneficial.

Ulcers of the cornea are so often a sign of bad health that we ought never to neglect such means as may be called for to improve the general condition.

Treating the matter clinically we shall find that the plan of local stimulation is most suitable for a large majority of the cases as they first come under notice, including phlyctenular cases, chronic super-ficial ulcers of various kinds, and even for many recent ulcers if not threatening suppuration. As a general rule, it is not applicable when there is much photophobia, but some well-marked exceptions to this rule are found, especially in old-standing cases. The most convenient remedy is the ointment of amorphous yellow oxide of mercury, in the proportion of from 1 to 3 grains to the drachm, and of this a piece about as large as a hemp-seed is to be put inside the eyelids once a day. If smarting continues more than about half an hour, or if the eye becomes distinctly more irritable after a few days' use, the ointment must be weakened or discontinued. Calomel flicked into the eye daily or less often is also an admirable remedy in the same cases. Nitrate of silver in the form of solid mitigated stick is useful if carefully applied to large conjunctival pustules, and occasionally to indolent corneal ulcers ; its use, how-ever, needs some practice, and it is seldom really necessary. Solutions of from 5 to 10 grains to the ounce are used by some surgeons instead of the yellow ointment, but require more skill and judgment. This remedy is particularly useful in old vascular ulcers and when there is sticky discharge.

Local stimulation combined with the soothing treatment by atropine gives good results in some cases where there is much photophobia and yet only superficial change. When in doubt it is best to depend for a few days on atropine, used often enough to cause wide dilatation of the pupil.

Severe photophobia, in young children especially, is sometimes beneficially treated by a free division of the outer canthus, by which the spasm is rendered impossible for a time, and remedies can be more efficiently used. In all cases of corneal disease attended with intolerance of light the patient is to wear a large shade over both eyes ; a little patch over one eye does not relieve photophobia. Many a child is kept within doors to the injury of its health who, if supplied with a suitable shade, would go out daily without the least detriment to its eyes.

In chronic and relapsing cases, with photophobia and irritability, where other methods have had a fair trial, a seton in the temple generally gives the best results, and this whether there is much congestion of the eye or not. A double thread of thick silk is used, and at least three quarters of an inch of skin included between the punctures, which are placed among the hair or behind the ear, so that the resulting scar may be hidden : it is to be moved daily, and if acting badly may be dressed with savin ointment. The seton should be worn at least six weeks. Severe inflammation, and even abscess, is sometimes set up within a few days of the insertion of the thread, and in very rare cases severe bleeding has occurred some days after, from ulceration of the coats of an arterial branch. To avoid wounding the artery in inserting the seton in the temple, the skin is to be held well away from the head.

Very severe recent phlyctenular cases are often very difficult to influence, and remain practically " blind " with spasm of the lids, for weeks. There is seldom any great risk, provided that we thoroughly examine the cornea at intervals of a few days, and they generally in the end make a very good recovery. Calomel dusted on the cornea sometimes helps more than any other local measure, and change of air, especially to the seaside, frequently effects a more rapid cure than any plan.

Cases for which the stimulating treatment is suitable seldom need the eye to be bandaged, though, as mentioned, they often need a shade.

The remaining methods of treatment—protective bandaging, atropine, hot fomentations, and operative measures—are reserved for the more serious forms of ulceration, the serpiginous ulcer, acute

suppurating ulcers, and abscess, and generally
for all ulcers where hypopyon is either present or
threatened, or where an ulcer is deep and threatens
to perforate. The compress used for this purpose
consists of a pad of cotton wool and a single turn of
bandage, tied at the back of the head. A piece of
linen rag, or of lint, may be placed between the skin
and the cotton wool. Such a compress is most grateful in the irritation caused by a corneal abrasion or
after a foreign body has been removed, a drop or two
of atropine being used if there be much congestion
or threatened iritis.

In these severe cases atropine is to be used regularly three or four times a day, on the ground that
iritis, if not present, is very likely to occur; it also
acts beneficially in soothing pain, and, perhaps, in
diminishing congestion. Hot fomentations are used
in the same cases. I generally direct that the compress be removed every two hours, or, sometimes,
every hour, and the lids fomented for fifteen or
twenty minutes with a belladonna lotion (one drachm
of extract to the pint) made as hot as can be
borne. Provided that atropine is used, there is
no actual need for the belladonna; hot water or
poppy-head fomentation is as good. But it is a good
routine to employ the belladonna, for the atropine is
not always used properly. A large number of cases of
acute suppurating ulcer, and of serpiginous ulcer,
and of abscess, will make quick recovery under this
treatment, combined with the administration of bark
and ammonia, and ether, or quinine. Even a considerable hypopyon will often be quickly absorbed.

But in some the ulceration increases, or the hypopyon, if present, enlarges. The hypopyon in such a
case is to be evacuated by an incision close to the
margin of the cornea. Many surgeons prefer at
the same time to make an iridectomy, but the real
effect of removing a piece of the iris on the progress
of the inflammation is doubtful. I incline to think

that a paracentesis with a broad needle, repeated if the hypopyon re-form in a few days, is all that is needed. Another method is to evacuate the aqueous by cutting across the whole width of the ulcer. and by opening the wound daily with a probe to keep the cornea flaccid until healing is well established; this method was intended by its author (Saemisch), especially for the serpiginous ulcers. In corneal abscess a similar incision is often made across the inflamed spot into the anterior chamber.

In all these operations the hypopyon, if present, will escape through the incision, and after all of them the anterior chamber will leak for a time. the length of which varies partly with the size of the incision. Very probably iridectomy is often so beneficial because the incision which is necessary does not close so soon as a mere puncture; I have several times seen the best results from a wound made with a keratome, as for iridectomy, but without the removal of any iris. When an ulcer is just about to perforate, as shown by protrusion of its thin transparent floor, puncture with a needle through the spot will aid the healing.

Very good results are seen from the use of cold evaporating lotions in some cases, chiefly, I think, in irritable superficial ulcers, with much congestion and spasm of lids.

Conical cornea.—In this condition the central part of the cornea very slowly bulges forwards, forming a bluntly conical curve. The focal length of the affected part of the cornea is thereby shortened, so that the eye becomes myopic, not by increase of its length but by increase of the refractive power of one of its media. The curvature, however, is seldom uniform, and hence irregular astigmatism generally complicates the myopia.

The disease, which is rare, occurs chiefly in young adults, especially women, and is generally dated from some illness or failure of general health;

and it appears to be due to defective nutrition of that part of the cornea which is furthest from the blood-vessels. In advanced cases the protrusion of the cornea is very evident, whether viewed from the front or from the side, but slight degrees are less easily distinguished from ordinary myopic astigmatism. In high degrees the apex of the cone may become nebulous. The disease may progress to a high degree, or stop before great damage has been done. Concave glasses alone are of little use, but in combination with a narrow slit or small central hole, in a screen by which the peripheral rays are cut off, benefit may be gained in the slighter cases. In advanced cases operation is needed (*see* Operations).

CHAPTER VIII

DIFFUSE KERATITIS

Syphilitic keratitis (interstitial, parenchymatous, or strumous keratitis)

In this disease the cornea throughout its whole structure undergoes a chronic inflammation, showing no tendency either to the formation of pus or to ulceration, the inflammatory products being, after several months, slowly absorbed either wholly or in great part, and the transparency of the cornea restored in proportion.

The visible changes in the cornea are usually preceded for a few days by some ciliary congestion and watering of the eye, and during this interval no diagnosis can be given. Then a faint cloudiness is seen in one or more large patches, and the surface, if carefully looked at, is found to be "steamy" (p. 73). These nebulous areas may be in any part of the cornea. In from about two to four weeks the whole cornea has usually passed into a condition of white haziness with steamy surface, of which the term "ground glass" gives the best idea. Even now, however, careful inspection, especially by focal light, will show that the opacity is by no means uniform, that it shows many whiter

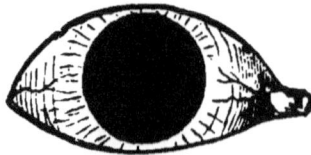

FIG. 17.

spots or large denser "clouds" scattered among the

general "mist;" in very severe cases the whole cornea is quite opaque and the iris entirely hidden; but, as a rule, the iris and pupil can be seen, though very imperfectly (Fig. 17). In many cases iritis takes place, and posterior synechiæ are formed. In

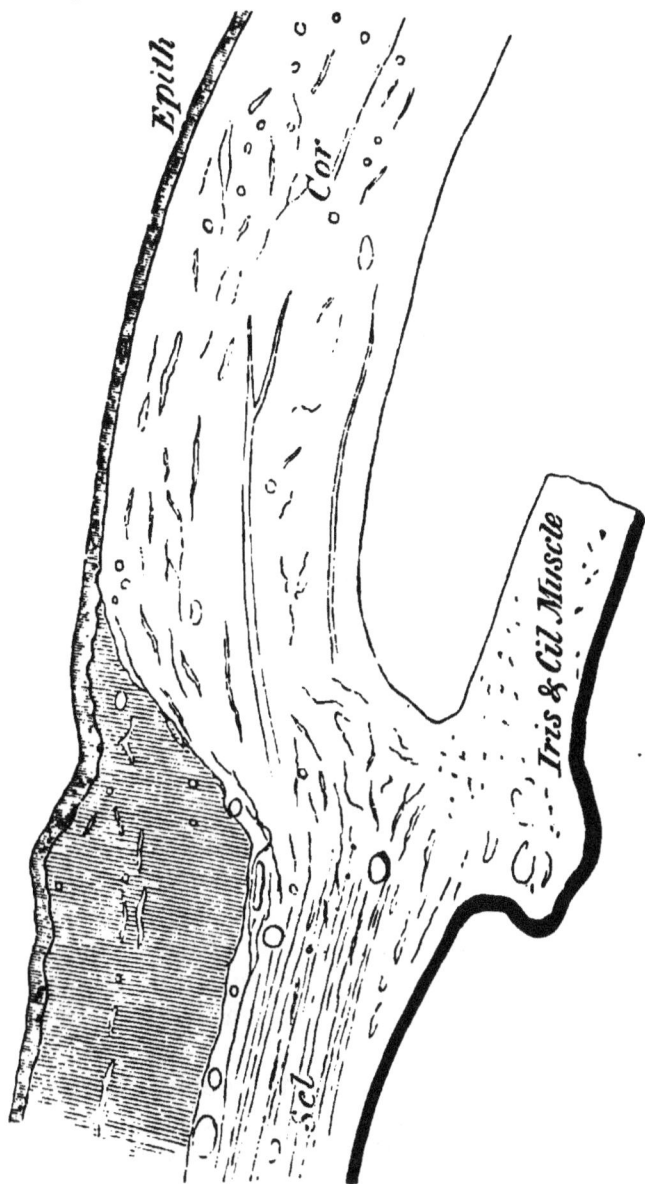

FIG. 18.—Thickening of cornea and formation of vessels in its layers in syphilitic keratitis. Subconjunctival tissue thickened.

the course of the keratitis blood-vessels are often formed in the layers of the cornea (Fig. 18). They are developed from branches of the ciliary vessels (Figs. 1 and 2); they are small and thickly set in patches, and, being covered by a certain thickness of hazy cornea, their bright scarlet is toned down to a dull reddish pink colour, the "salmon-patch" of Hutchinson. The separate vessels can be seen only if magnified, when we see that the trunks passing in from the border divide at acute angles into very numerous twigs, lying close to each other and taking a nearly straight course towards the centre (Fig. 19). These salmon-patches are of no constant shape, but when small are crescentic, and tend when large to the sector-shape. In another type the vascularity

FIG. 19.

begins as a narrow fringe of looped vessels contin-uous with the superficial loop-plexus of the corneal margin (Fig. 20, compare Figs. 1 and 2), and gra-dually extending from above and below towards the centre. The vessels in these cases are more super-ficial, and the corneal tissue in which they lie is always swollen by infiltration; they are described as "marginal keratitis" by some authors (compare p. 78). Nearly all the worst cases of this kind occur in syphilitic subjects, but I believe that some of these patients are at the same time strumous. In extreme cases of either type the whole cornea becomes vascular, or all excepting, perhaps, a central island.

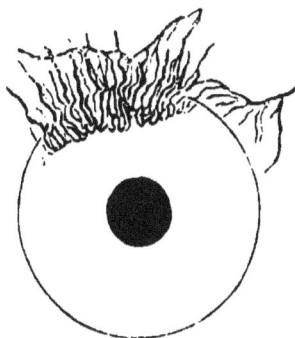

FIG. 20.

The degree of congestion of the eyeball, as well as the subjective symptoms, differ very much; as a general rule there is but moderate photophobia and pain, but when the ciliary congestion is great these symptoms are sometimes very severe and protracted.

The attack can be shortened and its severity lessened by treatment, especially by mercury; but the disease is always slow, and from six to twelve months may be taken as a fair average for its duration from beginning to end. In very bad cases with excessively dense opacity improvement sometimes continues for several years, and a very unexpected degree of sight is ultimately restored. Perfect recovery of transparency is less common, even in moderate cases, than is sometimes supposed, but the slight degree of haziness which so often remains does not much affect the sight. The epithelium regains its smoothness before the cornea becomes transparent; but at a late stage of a severe case some irregularities of surface and a few straggling vessels are sometimes seen, and the diagnosis may not at the first glance be quite easy.

Syphilitic keratitis is almost always symmetrical, though an interval of a few weeks commonly separates its onset in the two eyes; rarely the interval is several months, or even longer. It generally occurs between about the ages of 6 and 15; sometimes as early as 2½ or 3 years, and very rarely as late as 25 or 30. When it occurs at a very early age the attack is generally mild. Relapses of greater or less severity are common. Not only does iritis occur with tolerable frequency, but we occasionally meet with deep-seated inflammation in the ciliary region, giving rise either to secondary glaucoma or to stretching and elongation of the globe in the ciliary zone, or to softening and shrinking of the eyeball.* Dots of opacity are sometimes found on the back of the cornea early in the case and before the cornea itself is much altered (p. 94); sometimes, too, the interstitial exudation is much more dense at the lower part of the cornea than elsewhere. Syphilitic keratitis in strumous children often presents more irritability and photophobia and more conjunctival congestion than in others; but it is very seldom that ulceration occurs, and although in the worst cases the cornea becomes softened and yellowish, and for a time seems likely to give way, actual perforation and staphylomatous bulging are amongst the rarest events. In a very few cases, pannus from granular disease coexists with syphilitic keratitis.

Treatment.—A long but mild course of mercury exerts an undoubtedly good effect. It is customary to give iodide of potassium also, and it probably has some influence. If, as is often the case, the patients are very anæmic, iron, or the syrup of the iodide of iron, is sometimes a more useful adjunct to the mercury than iodide of potassium. Locally the use of atropine is advisable as a routine practice until the disease has reached its height, on the ground that iritis may be present. In cases attended by severe and prolonged photophobia and ciliary con-

---

* Patches of old choroiditis are often found after the corneæ have cleared. Probably in most of these the active stage of the choroiditis took place long before the keratitis set in.

gestion setons in the temples may give relief, and iridectomy is sometimes followed by rapid improvement; but the cases in which this operation is needed or justifiable are not numerous. When all inflammatory symptoms have subsided, the use of the yellow ointment or of calomel appears to aid the absorption of the residual opacity.

The form of keratitis above described is caused by *inherited* syphilis. In a few very rare cases it has been seen as the result of secondary *acquired* syphilis. Other cases of diffuse keratitis occur in which syphilis has no share, but they are seldom symmetrical, nor do they occur early in life. That diffuse chronic keratitis affecting both eyes of children and adolescents is, when well characterised, almost invariably the result of hereditary syphilis is proved by abundant evidence. A very large proportion of its subjects show in their own persons some of the other well-known signs of hereditary syphilis in the teeth, skin, ears (deafness), physiognomy, mouth, or bones. When the patients themselves show no other signs, careful questioning will often elicit the history of infantile syphilis in the patient or in some brothers or sisters, or of acquired syphilis in one or other parent. That this keratitis stands in no causal relation to struma is clear, because the ordinary signs of struma are not found oftener in its victims than in a corresponding number of children in ordinary health, because persons who are decidedly strumous do not suffer from this keratitis more often than others, because the forms of eye disease which are universally recognised as "strumous" (ophthalmia tarsi, phlyctenular disease, and relapsing ulcers of cornea) very seldom accompany this diffuse keratitis, and, further, because when such a coincidence exists, the case is noticed to differ from the ordinary type of interstitial keratitis by the addition of certain "strumous" symptoms, *e.g.* superficial vascularity, unusual irritability, and tendency to ulceration.

Other forms of keratitis

Inflammation of the cornea forms a more or less conspicuous feature in several diseases where the primary or principal seat of mischief is in some other part of the eye. It is important for purposes of diagnosis to compare these *secondary or complicating forms of keratitis* with the primary diseases of the cornea described in this and the preceding chapters.

In cases of iritis, the lower half of the cornea often becomes steamy, and its tissue more or less hazy. In other cases a number of small separate opaque dots are seen on the posterior elastic lamina (Descemet's membrane), often so minute as to need a hand lens for their detection. In other cases a few large dots only are present, or a mixture of large and small. They are sharply defined, the large ones looking very like minute drops of cold tallow, the smallest like grains of greyish sand ; in cases of long standing they may become either very white or highly pigmented. They are generally arranged in a triangle, with its apex towards the centre and its base at the lower margin of the cornea, and the smallest dots are commonly nearest the centre (Fig. 21), but in some

FIG. 21.

cases (sympathetic ophthalmitis especially) the dots are scattered over the whole area. They are of course difficult to detect in proportion as the cornea in front of them has lost its transparency.

The term *keratitis punctata* is used to express this accumulation of dots on the back of the cornea, and by some authors is made to include also small spots with hazy outlines, which lie in the cornea proper, and are sometimes seen in the same or in other cases. The condition is almost without exception secondary to some form of iritis or of inflammation of the choroid and vitreous. But a few cases are seen, chiefly in young adults, where the corneal dots form the principal if not the sole change. The number of these cases diminishes, however, in proportion to the care with which other lesions are sought.

It is now and then difficult to say whether the iritis or keratitis in a mixed case has been the initial lesion; but when this doubt arises the cornea has generally been the starting-point; and with care we are seldom at a loss to decide whether the case is one of syphilitic keratitis with iritis, or of cyclitis with corneal mischief and iritis, or of primary iritis with an unusual degree of corneal haze (see Cyclitis and Iritis).

Slight loss of transparency of the cornea occurs in most cases of *glaucoma*. The earliest change is a fine uniform steaminess of the epithelium. In very severe acute cases the cornea itself becomes hazy throughout, though never in a high degree. The same occurs in chronic cases of long standing with great increase of tension, but the epithelial "steaminess" often then gives place to a coarser "pitting," with little depressions and elevations, especially on the part which is ordinarily uncovered by the lids.

A peculiar and rare form of corneal disease is the *transverse calcareous film*, an elongated patch of light grey opacity, looking when magnified like very fine sand, placed beneath the epithelium and running almost horizontally across the cornea. It consists of minute crystals, chiefly calcareous, is seen in elderly or prematurely senile persons, and probably stands in some relation to the uric-acid diathesis.

Arcus senilis is caused by fatty degeneration of the corneal tissue just within its margin. It generally begins beneath the upper lid, and next appears beneath the lower, forming two narrow, white or yellowish crescents, the horns of which finally meet at the sides of the cornea; it always begins and remains most intense on a line slightly within the sclero-corneal junction, and the degeneration is most marked in the superficial layers of the cornea beneath the anterior elastic lamina; in other words, the change is greatest at the part most influenced by the marginal blood-vessels. Its presence is not found to interfere with the union of a wound carried through the diseased tract. Nevertheless, its occurrence chiefly at advanced ages, and its frequent coexistence with fatty degeneration, not only in distant parts of the body, but in the blood-vessels and muscles of the eyeball, mark it as an essentially senile change.

Less regular forms of arcus are seen as the result of prolonged or relapsing inflammations near the corneal border, whether ulcerative or not. It is generally easy to distinguish such an arcus, because the opacity is more patchy and its outline less regular than in the primary form; when arcus is seen unusually early in life it is generally of this inflammatory kind, for simple arcus is comparatively rare below forty.

Opacity of a very characteristic kind is likely to follow the use of a lotion containing *lead* when the surface of the cornea is abraded. An insoluble, densely opaque and very white film of lead salts is precipitated on the ulcerated surface, and adheres very firmly to it. Such an opacity when once seen can scarcely be mistaken; it is sharply defined, and as white as white paint. If precipitated on a deep and much inflamed ulcer, the film of tissue to which it adheres is often thrown off, but in the common cases, where the corneal disease has been a superficial abrasion or ulcer, the lead adheres very firmly, and can only be scraped off imperfectly, and with great difficulty. But even in these cases the layer is probably after a time thrown off or worn off, if we may judge by the fact that nearly all the lead opacities which come under notice are comparatively new. The practical lesson is, of course, never to use a lead lotion for the eye when there is the least suspicion that the corneal surface is broken. Powdered acetate of lead rubbed into the conjunctiva (a treatment sometimes used for granular lids) is, I believe, not attended by risk of corneal opacity, even though there be ulceration; the lead is precipitated at once, and adheres for weeks to the surface of the granular conjunctiva, any superfluous salt being washed away with water immediately after the powder has been applied.

The prolonged use of *nitrate of silver*, whether in a weak or strong form, is sometimes followed by a dull (brownish-green) permanent discoloration of the conjunctiva, and even the cornea may become slightly stained.

CHAPTER IX

INFLAMMATION of the iris may be caused by a specific blood disease, as syphilis; or may be the expression of a tendency to relapses of inflammation in certain tissues under the influence largely of climate and weather, rheumatic iritis; or may occur with ulcers of the cornea, or in the course of an inflammation set up by local causes in the cornea, as after corneal abrasions and operation wounds, particularly for extraction of cataract; finally, iritis forms a very important part of the grave and peculiar disease known as sympathetic ophthalmitis.

Acute iritis, from whatever cause, is shown by a change in the colour of the iris, by indistinctness of its texture ("muddiness"), by diminution of its mobility, and by the existence of adhesions (*posterior synechiæ*) between its posterior (uveal) surface and the capsule of the lens; there is, besides, in most cases, a certain dulness of the whole iris and pupil, caused partly by slight corneal changes (p. 94), partly by muddiness of the aqueous humour. The eyeball is congested and sight is almost always defective. There may or may not be pain, photophobia, and lachrymation.

The congestion is often nearly confined to a zone of about one twelfth or one eighth of an inch wide, which surrounds the cornea, its colour being pink (not raw red), the vessels small, taking a radial or nearly straight course, and lying beneath the conjunctiva (ciliary congestion). These are the episcleral branches of the anterior ciliary arteries (pp. 13 and 75). Quite the same congestion is seen in many other conditions, *e.g.* corneal ulceration; whilst on the other hand, in some cases of iritis, the superficial (conjunctival) vessels are congested also, especially in their anterior division, which is chiefly an offshoot of the ciliary system (Figs. 1 and 2). We therefore never diagnose iritis from the character of the congestion alone; but iritis being proved by the other symptoms (see below), the kind and degree of congestion help us to judge of its severity.

The altered colour of the iris is explained by its congestion, and by the effusion of lymph and serum into its substance; a blue or grey iris becomes greenish, whilst a rich brown one is but little changed. The inflammatory swelling of the iris also accounts both for the blurring (muddiness) of its beautifully reticulated structure and for the sluggishness of movement caused by stiffness of its tissue, noticed in the early period. After a few days lymph is thrown out at one or more spots on the posterior surface, which, by adhering to the lens capsule, still further hampers its movements; and most cases do not come to notice till some such synechiæ have formed. The quantity of solid exudation, both on the hinder surface and into the structure of the iris, is open to much variation, and is usually greatest in syphilitic iritis, when distinct nodules of pink or yellowish colour are sometimes seen projecting from the front surface. In rare cases pus is thrown out by the iris into the aqueous, and, sinking down, forms a hypopyon. Lastly, firm adhesions to the lens capsule may be present without much evidence of exudation into the structure of the iris. All these exudative changes are most abundant at the inner ring of the iris, where its capillary vessels are far the most numerous (Fig. 22).

The dulness and apparent discoloration of the iris are also due, in part, to turbidity of the aqueous from suspension in it of pus or blood-corpuscles, either of which may form a distinct deposit at the lowest part of the anterior chamber (hypopyon, hyphæma). Sometimes the slightly turbid fluid

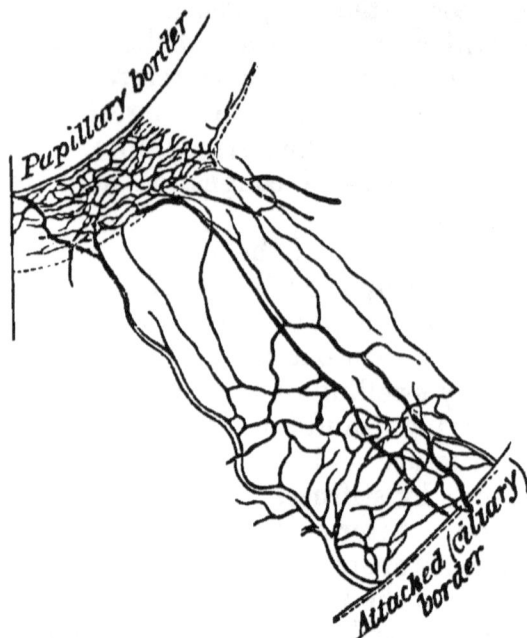

FIG. 22.—Vessels of human iris; capillaries most numerous at pupillary border, and next at ciliary border.

coagulates into a gelatinous mass, which almost fills the chamber.

The tension of the eyeball is often a little increased in acute iritis ; rarely it is considerably diminished, but in such cases there are generally other pecu- liarities.

The condition of the pupil alone is diagnostic in all except very mild or incipient cases of iritis. It will be sluggish or quite inactive, and not quite round ; it will also be rather smaller than its fellow (supposing the iritis to be one-sided), because the surface of the iris is increased (and the pupil, there- fore, encroached on) whenever its vessels are dis- tended (p. 15). Atropine will cause it to dilate between the synechiæ. These synechiæ, being fixed appear as angular projections when the iris on each

:side of them has retreated. If there be only one adhesion it will merely notch the pupil at one spot; if they be numerous the pupil will be crenated or irregular (Fig. 23). If the whole pupillary ring, or

FIG. 23.—Posterior synechiæ causing irregularity of pupil.

still more, if the entire posterior surface of the iris be adherent, scarcely any dilatation will be effected: the former condition is called annular or circular synechia and its result is "*exclusion*" *of the pupil;* the latter is known as *total posterior synechia.* If the synechiæ are new and the lymph soft the repeated use of atropine will cause them to give way, and the pupil will become round, but even then some of the uveal pigment, which is easily separable from the posterior surface of the iris, will remain behind on the lens capsule, glued to it by a little lymph (Fig. 24); and the presence of one or more such

FIG. 24.—Spots of pigment and lymph at seat of former adhesions.

spots of brown pigment on the capsule is always as good proof of present or of past iritis as the existence of the synechiæ themselves. The pupillary area itself is often blurred or even quite obscured by lymph, which spreads over it from the iris, and will then be greyish or yellowish according to the quantity. The iris may be inflamed without any lymph being effused from its hinder surface, and then the pupil, though sluggish and acting imperfectly to atropine, being, perhaps, rather oval and never dilating widely, will present no angular extensions, nor any adhesion of separate pigment to the lens, but it will always be discoloured (serous iritis); iritis of this kind often occurs with ulceration of the cornea. When exudation into the pupil becomes organised a dense white membrane, or a delicate film, often however presenting one or more little clear holes, is formed over the pupil ("*occlusion*" *of the pupil*).

Pain referred to the eyeball and to the parts supplied by the first, and sometimes by the second, division of the fifth nerve is a common accompaniment of iritis, especially in the early period of the attack. It is a very variable symptom, and gives no clue to the amount of structural change going on in the parts, being sometimes quite an insignificant feature in a case where much lymph is thrown out. The pain is seldom constant, but comes on at regular or irregular intervals daily, and is described as shooting or throbbing or aching. It is commonly referred to the temple or forehead, as well as to the eyeball, but sometimes to the side of the nose and to the upper teeth. Photophobia and watering of the eye are generally proportionate to the pain.

The duration of acute severe iritis varies from a few days to many weeks. The defect of sight is proportionate to the haziness of the cornea, aqueous, and pupillary space, and in some cases is increased by changes in the vitreous. In some cases, iritis sets in very gradually, causing no marked congestion or pain, but slowly giving rise to the formation of tough adhesions, and often to the growth of a thin membrane over the pupillary area. In some of these the iris becomes thickened and tough, and its large vessels undergo much dilatation, and in others keratitis punctata occurs (see Cyclitis, p. 113; diseases of cornea, p. 94; and sympathetic ophthalmitis, p. 117). As would be expected, there are all degrees of acuteness and severity.

Results of iritis.—Such of the results as are per-manent need separate notice. Reference has been made to the adhesions, which are often permanent, and to the spots of uveal pigment on the lens capsule, which are always so. Either of these conditions tells a tale of past iritis, which is often a valuable aid to diagnosis. A blue iris which has undergone inflam-mation may remain permanently greenish.

When the pupil is "excluded" or "occluded," the remainder of the iris being free, fluid is apt to collect in the posterior chamber, and by bulging the iris forwards, and diminishing the depth of the anterior chamber, excepting at its centre, to give to the pupil a funnel shape. If such bulging be partial, or be divided by bands of tough membrane, a cystic ap-pearance is given. *Secondary glaucoma* is likely to follow, and the tension of the globe should, there-fore, be carefully noted whenever this bulging is pre-sent. "Total posterior synechia" always shows a severe iritis, though often one of chronic course; it often signifies deep-seated disease, and may be complicated by secondary opacity of the lens (cata-ract). Relapses of iritis are believed to be induced by the presence of synechiæ, even when there is no protrusion of the iris by fluid; but their influence in this way has probably been much overrated.

The following are the most important points as to the causes of iritis and the chief clinical differences between the several forms.

Constitutional causes. — Syphilis. — The iritis is acute; there is a great tendency to effusion of lymph and formation of vascular nodules (plastic iritis), and the nodules, when very large, may even suppurate. The disease is very often symmetrical, but want of symmetry and absence of lymph nodules, are common. It occurs only in secondary syphilis (either acquired or inherited), and seldom relapses. It is to be carefully distinguished from the iritis complicating syphilitic keratitis (p. 89).

Rheumatism is the cause of most cases of relapsing
unsymmetrical iritis ; there is little tendency to effusion of lymph, and nodules are never formed, but
there is sometimes fluid hypopyon (p. 80) ; the congestion and pain are often more severe than in
syphilitic iritis. A single attack is very rarely symmetrical, though both eyes commonly suffer by turns.
It relapses at intervals of months or years. Even repeated attacks sometimes result in but little damage
to sight.   *Gout* is apparently a cause of some cases
of both acute and insidious chronic iritis. It is perhaps doubtful whether the gout or the chronic rheumatism to which many of the same patients are
liable is the cause of the iritis. In its tendency to
relapse and to affect only one eye at a time gouty
resembles rheumatic iritis. The children of gouty
parents are occasionally liable to a very insidious and
destructive form of chronic iritis, with disease of the
vitreous (see Chapter on Etiology), keratitis punctata,
and glaucoma. *Smallpox* is occasionally followed by
simple acute iritis.

A few cases of symmetrical, chronic, progressive
iritis, complicated with choroiditis, disease of vitreous
and secondary cataract, are seen, for which it is at
present impossible to assign any cause either general
or local. They are chiefly seen in young adults,
and, I think, oftenest in women.

Sympathetic iritis.—See Sympathetic Ophthalmitis.

Local causes.—Injuries.—Perforating wounds of
the eyeball are often followed by iritis, particularly
if the wound be irregular and contused, and complicated with wound of the lens. Perforating wounds
are more likely to be followed by iritis in old than in
young persons. If the corneal wound suppurate or
become much infiltrated the iritis is likely to be
suppurative. Iritis may follow a wound of the lens-
capsule without wound of the iris, and with only a
mere puncture of the cornea. Examples of all these
forms of traumatic iritis are seen after the various

operations for cataract. The iritis following extraction of senile cataract is often prolonged, attended by chemosis, much congestion, and the formation of tough membrane behind the iris. It will be described in more detail under " Cataract." Iritis may follow superficial wounds, or abrasions of the cornea, or direct blows on the eye. It is also a frequent accompaniment of ulcers of the cornea especially when deep, or complicated with hypopyon, or occurring in elderly persons. It is of great importance in every case of iritis where the question of injury comes in, to ascertain whether or not there has been a perforating wound. Iritis may be secondary to deep-seated disease or tumour in the eye.

Treatment.—(1.) In every case where iritis is present atropine is to be used often and continuously, in order to break down adhesions which have formed, and to allow any lymph subsequently thrown out to be situated outside the area of the ordinary pupil. A strong solution (four grains of sulphate of atropine to one ounce of distilled water) is to be dropped into the conjunctival sac every two hours in the early period. In many cases the synechiæ are, when first seen, already so tough that the atropine has no effect on them; but even then it may still prevent new ones forming on the same circle. Moreover, the pupil when kept as widely dilated as possible is less likely to be covered over by lymph or organised membrane from the iris than if contracted. Atropine also undoubtedly relieves the pain and congestion in iritis.

(2.) If there is severe pain with much congestion three or four leeches should at once be applied to the temple, to the malar eminence, or to the side of the nose ; some surgeons prefer the septum nasi. They may be repeated daily in the same or a smaller number with advantage for several days, if the symptoms are severe ; or after one leeching repeated blistering may be substituted. Some surgeons use opiates instead of, or in addition to, leeches.

(3.) The further treatment must depend mainly on the cause of the disease, into which careful inquiry should always be made. If the iritis be syphilitic, treatment for secondary syphilis is proper, mercury being given to very slight salivation for several months, even though all the active eye symptoms quickly pass off. The rheumatic and gouty varieties are less definitely under the influence of internal remedies; iodide of potassium, alkalies, and colchicum certainly appear to exert a good effect in some cases; when the pain is severe tincture of aconite is sometimes markedly useful; mercury is seldom needed, but in protracted and severe cases may sometimes be used with advantage.

(4.) Rest of the eyes is a very important curative measure. Many a case is lengthened out and many a relapse after partial cure is brought on by the patient continuing at work, or beginning to use his eyes too soon. It is not in most cases necessary to remain in a perfectly dark room ; to wear a shade in a room with the blinds down is generally enough, provided that no attempt is made to use the eyes. Work should not be resumed till at least a week after all congestion has gone off.

(5.) Avoidance of cold draughts of air on the eye and of all causes of " catching cold " is also very important. It is best to keep the eye warmly tied up with a large pad of cotton wool especially in rheumatic iritis.

(6.) As a rule no stimulants are to be allowed, and the bowels to be kept well open. It is sometimes advisable to combine quinine with the mercury in syphilis, or to give it in addition to other remedies in rheumatic cases.

(7.) Iridectomy is needed for cases of severe iritis where judicious local and internal treatment have been carefully tried for some weeks without marked relief to the inflammatory symptoms, whether or not there be increased tension. It is chiefly in cases of constitutional origin, either syphilitic or rheumatic, that it is necessary, and in the iritis accompanying ulcers of the cornea. It is not applicable to sympathetic iritis, nor to iritis after cataract extraction, and comparatively seldom to other kinds of traumatic iritis.

In reference to iridectomy it is to be borne in mind that unless necessary it is injurious, by producing an enlarged and irregular pupil through which the patient will often not see so well as through the natural pupil, even though this be partially obstructed. The effect of the operation in staying and abating the inflammation is very marked in some cases, but in order to be sure that the effect is due to the operation we must have first tried fairly the other means of cure. Lastly, in regard to all methods of treatment we must bear in mind that acute iritis occurs in all degrees of severity, and that the mild cases often need little more than atropine, warmth, and rest, though more active medication is generally advisable on prophylactic grounds.

Traumatic iritis, especially in the early stage, is often best combatted by continuous cold applied by means of pieces of lint wetted in iced water or on a block of ice and laid upon the lids; and by leeches. Cold is not applicable to any of the other forms of iritis.

Congenital irideremia (absence of iris) is occasionally seen, and is often associated with other congenital defects of the eye.

Coloboma of the iris (congenital cleft in the iris) gives the effect of a very regularly made iridectomy. It is always downwards or slightly down-in, and usually, but not always, symmetrical. There are many varieties in degree, and sometimes there is nothing more than a sort of line or seam in the iris. It may occur without coloboma of the choroid.

Persistent *remains of the pupillary membrane* have sometimes to be distinguished from iritic adhesions. They occur in the form of thin shreds or loops of tissue, resembling the iris in colour, and attached to its *anterior* surface close to the pupillary border. They are longer and slenderer than posterior synechiæ, and are often quite unattached to the lens-capsule. In one remarkable case I saw well-marked remains of this membrane complicated with equally unequivocal iritic adhesions in acute iritis in a man.

CHAPTER X

THIS chapter is intended to include cases of which the ciliary body itself, or the corresponding part of the sclerotic, or the episcleral tissue is the sole seat, or at least the headquarters, of inflammation.

From the richness of its vascular and nervous supply and its nutritive relations to the parts in its neighbourhood, we are not surprised to find that many of the morbid processes of this region show a strong tendency to spread, according to their precise position and depth, to the cornea, or iris, or vitreous, and by influencing the nutrition of the lens to cause secondary cataract.

Although alike on pathological and clinical grounds it is necessary to subdivide the class into groups, we may observe that in some of their more obvious and important characters all the diseases of this part show a general agreement; thus all of them are liable to relapse, and in all there is a marked tendency to patchiness, the morbid process being most intense in certain spots of the ciliary zone, or even occurring in quite discrete patches. It is convenient to make three principal clinical groups, the differences between which are accounted for to a great extent by the depth of the tissue chiefly implicated. The most superficial may be taken first.

(1.) *Episcleritis* is the name given to a large patch or patches of congestion, with some elevation of the conjunctiva, caused by thickening of the subjacent tissues, and situated in the ciliary region. The congestion generally affects the conjunctival as well as the deeper vessels, and the yellowish colour of the exudation tones the bright blood-red down to a more or less rusty tinge, which is especially striking at the centre of the patch. The thickening is greatest at the centre; it varies in amount, but seldom causes more than a low widely spread mound of swelling.

Episcleritis is most common in the exposed parts of the ciliary region, and especially near the outer canthus, but the patches may occur at any part of the circle; and exceptionally the inflammation is diffused over a much wider area than the ciliary zone, extending back out of view. No iritis occurs, but the iris may be a little discoloured and the pupil sluggish. The patient often complains of much aching pain. The disease is subacute, reaching its acme in not less than two or three weeks, whilst often a much longer time is required before absorption is complete. Fresh patches are apt to spring up while old ones are declining, and so the disease may last for months; indeed, relapses sooner or later (in fresh spots) are the rule in this disease. It usually affects only one eye at a time, but both often suffer sooner or later. After the congestion and thickening have disappeared a patch of the underlying sclerotic, of rather smaller size, is generally seen to be dusky as if stained; it is doubtful whether such patches represent thinning of the sclerotic from atrophy, or only staining; it is but seldom that they show any tendency to protrude.

In rare cases the exudation is much more abundant, and a large hemispherical swelling is formed, which may even contain pus; such cases pass by gradations into conjunctival phlyctenulæ, and are generally seen in children (compare p. 78).

Episcleritis of the common type is seldom seen excepting in adults, and is commoner in men than in women. Not only is it commonest on the exposed parts of the globe, but inquiry often shows that the sufferer is, either from occupation or temperament, particularly liable to be affected by exposure to cold or to changes of temperature; some are decidedly rheumatic. Nearly similar patches are occasionally seen as the result of tertiary syphilis, acquired or inherited.

In the treatment protection by a warm bandage, rest of the eyes, local stimulation of the swelling by the yellow ointment, and the use of repeated blisters, are generally the most efficacious. Atropine is often useful in allaying the pain. Internal remedies seldom seem to exert much influence; iodide of potassium appears to do good in some cases, and quinine in others; tincture of aconite is sometimes given with apparent advantage.

(2.) *Cyclo-keratitis and cyclo-iritis* (scrofulous sclerotitis, anterior staphyloma).—A very persistent and relapsing subacute inflammation, characterised by the congestion having a violet tinge (deep scleral congestion, p. 13), being abruptly limited to the ciliary zone, and affecting some parts of the zone more than others (tendency to patchiness). Early in the case there is a slight degree of bulging of the affected part, due partly to thickening; whilst patches of cloudy opacity, which may or may not ulcerate, appear in the cornea close to its margin. Later on iritis generally occurs. Pain and photophobia are prominent symptoms. After a varying interval, but seldom sooner than several weeks, all the symptoms recede. At the focus of greatest congestion, or it may be around the entire zone, the sclerotic is left of a dusky colour, sometimes interspersed with little yellowish patches representing imperfect absorption, and with permanent haziness of the most affected parts of the cornea. The disease is, however, almost certain to relapse sooner or later, and often a succession of fresh inflammatory foci follow each other without any intervals of real recovery, the whole process extending over many months. After each attack more haze of cornea and fresh iritic adhesions are left. The sclerotic, in bad cases of some years' standing, becomes much stained, and bulges very considerably (ciliary or anterior staphyloma), and the cornea becomes both opaque aud altered in curve; the eye is then useless, though but seldom liable to further active symptoms.

The characteristic appearance of an eye which has been moderately affected is the dusky colour of the sclerotic and the notched or irregular boundary of the cornea, many of the opacities being continuous with the sclerotic (Fig. 25). The disease does not

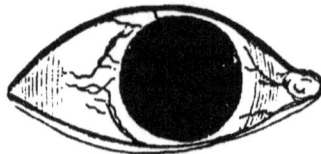

FIG. 25.

occur in children, nor does it begin late in life; most of the patients are young or middle-aged adults, and a large majority women. It is not associated with any special diathesis or dyscrasia, but generally goes along with a feeble circulation and liability to "catch cold;" in some cases there is a definite family history of scrofula or of phthisis. The predisposed are more likely to suffer in cold weather, or after change to a colder or damper climate, and any cause of exhaustion, such as suckling, will greatly increase the liability.

Treatment is at best but palliative. Local stimulation by yellow ointment or calomel is very useful in some cases, particularly those which verge towards the phlyctenular type. In the early stages, especially when the congestion is very violet, and altogether subconjunctival, atropine often gives relief, and it is, of course, useful for the iritis. Repeated blistering is also to be tried, though not all cases are benefited by it. I have not seen much benefit from setons. Warm, dry applications to the lids are, as a rule, better than cold. Mercury, in small and long-continued doses, is certainly valuable when the patient

is not anæmic and feeble, but it is to be combined
with cod-liver oil and iron.   Local protection from
cold and bright light by " goggles" is very important
during the attack and as a precaution against relapses
in cold weather.   There is no rule as to symmetry; both eyes often suffer sooner or later, but
sometimes one escapes whilst the other is attacked
repeatedly.

(3) *Cyclitis with disease of vitreous and keratitis punctata* (chronic irido-choroiditis).—A small but important series of cases, in which, with congestion like
that attending mild iritis, and with dulness of sight,
but with little or no pain or photophobia, flocculi are
found in the anterior part of the vitreous, or numerous small dots of deposit are formed on the posterior surface of the cornea (keratitis punctata, see
Fig. 21), the aqueous is often discoloured, and
insidious iritis usually follows.   Persistence and
liability to relapse are features as marked here as
in the other members of the cyclitic group, the final
condition turning very much on the extent of the
iritic adhesions, for when the synechiæ are numerous and tough, and the iris much altered in structure, secondary glaucoma may arise or the pupil be
blocked by iritic membrane.   When seen quite early
such a case will probably be diagnosed as " serous
iritis" or as "ciliary congestion," unless carefully
examined, the pupil being generally free in all parts,
or showing, at most, one or two adhesions when
atropine is used; glaucomatous symptoms, however,
sometimes develope early in the disease, before iritic
adhesions have formed.   In a few cases the punctate
deposits on the back of the cornea constitute almost
the only objective change (simple keratitis punctata),
but these are very rare.

The cases occur always in adolescents or young
adults, and the disease is always sooner or later symmetrical.   Many mild cases recover perfectly, and in

others a good result is finally achieved. In respect to cause, there is strong reason to believe that many of these cases are the result of gout in a previous generation, the patient himself never having had the disease (Hutchinson). The disease seems often to be excited in predisposed persons by prolonged overwork or anxiety, combined with underfeeding, or, what comes to the same thing, defective assimilation; the patients often describe themselves as, or are obviously, delicate. On the other hand, in some of the most distinctive cases, leading to secondary cataract and ultimately to shrinking of the eyes, the patient appears to be, from first to last, in good health, and free from any ascertainable morbid diathesis.

In the *treatment* prolonged use of atropine and complete, or at least relative, rest of the eyes, are the most important local measures. In certain cases iridectomy is necessary. Small doses of iodide of potassium and mercury appear to be useful in the earlier stages, given with proper precautions, and accompanied by iron, cod-liver oil, and sometimes quinine or bitters. In the worst cases, where the changes are very like those resulting from sympathetic ophthalmitis, no treatment seems to have much effect.

Cases of acute inflammation are occasionally seen in which most of the symptoms resemble those of acute iritis, but the iris is so little affected as to be evidently not the headquarters of the morbid action. The tension may be much reduced, whilst repeated and rapid variations, both in sight and objective symptoms, occur. The term "*idiopathic phthisis bulbi*" has been applied to some of these. Again, some cases of syphilitic inflammation, which are classed as syphilitic "iritis," might more correctly be called "cyclitis." In some cases of heredito-syphilitic keratitis there is much cyclitic complication (p. 91), and these are always difficult to treat.

Plastic or, more rarely, purulent inflammation of the ciliary body, following injury, is the usual starting-point of the changes which set up sympathetic inflammation of the fellow eye; and the changes in the sympathising eye generally begin also in the ciliary body quickly spreading forwards to the iris, and backwards to the choroid, vitreous and retina.

The outset of traumatic cyclitis is signalised by ciliary congestion, pain, and marked tenderness to palpation; there is often lowered tension and iritis. If the lens is transparent a yellow or greenish reflexion is, after a few days, often seen from behind it, and indicates the presence of lymph or pus spreading from the ciliary body into the vitreous.

SYMPATHETIC IRRITATION, AND SYMPATHETIC OPHTHALMITIS

Certain morbid changes in one eye may set up functional disturbance and destructive inflammation in its fellow. The term *sympathetic irritation* is given to the former, and *sympathetic ophthalmitis* (or *ophthalmia*) to the latter, and it is very important to distinguish between them.

Although at present the exact nature of the process which causes sympathetic inflammation is unknown, and though its path has not been fully traced out, it is certain (1) that the change starts from the region most richly supplied by branches of the ciliary nerves (composed of fibres from the fifth, sympathetic, and third), viz. the ciliary body and iris; (2) that its first effects are nearly always seen in the same part of the sympathising eye; (3) that in the exciting eye inflammatory or cicatricial changes are always present, such as are capable of setting up inflammation in the ciliary nerves; (4) that inflammatory changes have in some cases been found in the ciliary nerves of the exciting eye; (5) neither the optic nerves nor the blood-vessels are now believed to have any share in the transmission, and it is considered certain that the morbid influence must pass along the ciliary nerves, perhaps as neuritis, ascending to some centre and thence descending to the other eye.

In a large majority of the cases where sympathetic mischief occurs it is set up by a perforating wound (either accidental or operative) in the ciliary region of the other eye. The ciliary region is the zone, about an eighth of an inch wide, surrounding the cornea. The risk attending a wound in this " dangerous zone" is increased if it be lacerated, or heal slowly, or if the iris or ciliary body are engaged between the lips of the sclerotic, or if the eye contain a foreign body; under all conditions, indeed, which make the occurrence of plastic or purulent cyclitis probable. But sympathetic disease may follow any form of spontaneous chronic or acute inflammation which leads to the formation of organised tissue on the ciliary body (plastic irido-cyclitis), or to shrinking of the eye and cicatricial pressure or traction on the ciliary nerves. It may be caused by purulent ophthalmitis; or by the irido-cyclitis which may follow dislocation of the lens, or occur during the growth of an intra-ocular tumour. A foreign body lodged in the eye, whether the wound be in the ciliary region or not, is always a possible source of sympathetic mischief; and a wound entirely corneal, if accompanied by a large anterior synechia and dragging on the ciliary body, may also be a cause.

Symptoms in the exciting eye.—The exciting eye generally shows ciliary congestion and photophobia, and often suffers neuralgic pain when it is causing sympathetic *irritation.* Well-marked symptoms of cyclitis (p. 115), with tenderness on pressure, may be present in an eye which is causing sympathetic ophthalmitis ; but these symptoms may be so trivial that the patient overlooks them, and so transient as to have passed off before he is seen, and yet destructive changes be set up in the sympathising eye. It is especially important to remember that the exciting eye need not be blind, the exciting change being localised in a small part of its anterior portion, and that, under certain circumstances, it may in the end be the better eye of the two.

Symptoms of sympathetic irritation in the sympathising eye.—The eye is, in common speech, "weak" or "irritable." It is intolerant of light, and easily flushes up, and becomes watery if exposed to bright light or if much used ; the accommodation is weakened or irritable, so that continued vision for near objects is painful or even impossible, and the ciliary muscle seems liable to give way suddenly for a short time, the patient complaining that near objects now and then suddenly become misty for a while. Temporary darkening of sight, indicating suspension of retinal function, is sometimes observed, whilst in other cases we find considerable lasting defect of sight without ophthalmoscopic changes, and the cause of which is obscure. Neuralgic pains referred to the eye and side of the head are also common. Such attacks may occur again and again in varying severity, lasting for days or weeks, and finally ceasing without ever passing on to structural change. They cease quickly on removal of the offending eye.

Symptoms and course of sympathetic inflammation (ophthalmitis).—The disease may arise out of an attack of "irritation," but more commonly sets in without any such warning. It may be acute and severe, or so insidious as to escape the notice of the patient until well advanced. It is in all cases a prolonged and a relapsing disease, and when once started is self-maintaining, its course extending over many months or even a year or two. In mild cases a good recovery eventually takes place, but in a large majority the eye becomes practically, if not absolutely, blind. The disease is essentially an irido-cyclitis or irido-choroiditis, the external signs being those of iritis with rapid formation of tough and extensive synechiæ. Its chief early peculiarities are a great tendency to dotted deposits on the back of the cornea (p. 94), a dusky tint of ciliary congestion with marked engorgement of the large perforating vessels which pass through the sclerotic in the ciliary region (as in glaucoma), and to marked thickening and muddiness of the iris, the anterior chamber becoming shallow; we must add that there is frequently tenderness on pressure in the ciliary region. If the pupil allows of ophthalmoscopic examination we shall find the vitreous clouded by floating opacities, and there may be neuro-retinitis. In acute and severe cases the congestion is intense, there is severe pain and tenderness on pressure, and the iris, besides being thick, is changed in colour to a peculiar buff or yellowish brown, and shows numerous enlarged blood-vessels ("plastic" form). In cases of all degrees the tension is often increased, the eye becoming decidedly glaucomatous for a longer or shorter time. The lens often suffers, showing many small dotted opacities, and eventually becoming more or less completely opaque. In the worst cases the eye finally shrinks, but in many a permanently glaucomatous state is established with slight thinning and bulging of the sclerotic in front, total posterior synechia, and secondary cataract. In the mildest cases (the so-called "serous" form), the disease never goes beyond a chronic iritis with punctate keratitis and disease of the vitreous.

Sympathetic ophthalmitis generally begins about two or three months after the injury or other cause of mischief in the exciting eye, seldom sooner than three weeks, i. e. not until time has elapsed for well-marked inflammatory changes to occur. Cases are on record at a much shorter interval, but they are

excessively rare; on the other hand, the disease may
set in at any length of time, even many years after
the injury or other disease of the exciting eye, particularly if the latter has remained liable to attacks of
irritation, or if it contain a foreign body, or if ossifi-
cation of ᵼorganised exudations in the choroid or
ciliary body has taken place. It occurs at all ages,
but children are considered to be more liable than
adults.   Definite inflammatory symptoms are generally present in the exciting eye, but as already stated
they are sometimes very slight.   These symptoms in
the exciting eye always precede by some days the
commencement of structural disease in the sympathising eye; the morbid process seems to take some
days to reach the second eye.

Treatment.—By far the most important measure
refers to prevention. When once inflammation has
begun we can do comparatively little to modify its
course. The clear recognition of this fact leads us
to advise the excision of every eye which is at the
same time useless and liable to cause sympathetic
mischief, *i. e.* of all eyes which are blind from disease
of the anterior segment of the globe; and to give this
advice most urgently when the blind eye is already
tender or irritable, or is known to be liable to become
so, when it has been lost by wound, and when it is probable that it may contain a foreign body. Any lost
eye in which there are signs of past iritis, whether or
not it has been injured, is best removed, especially
if shrunken. But much judgment is needed if the
damaged eye, though irritable and likely to cause
mischief, still has more or less sight. Every attention must be paid to the exact position of the wound,
the evidence as to its depth, the condition of the
lens, the evidence of hæmorrhage, and especially of
that white or yellowish haziness (lymph) behind the
lens which implies cyclitis. The interval since the
injury and the condition of the wound, whether
healed by immediate union, or with scarring or

puckering or flattening, are very important points. *Irritation* of the fellow eye may set in a few days after the injury; but since *inflammation* very seldom begins sooner than two or three weeks, if we see the case early we may watch it for a little time. Complete and prolonged rest in a darkened room is a very important element in the prevention of sympathetic irritation and inflammation, and should always be insisted on when we are trying to save an injured eye (compare pp. 107 and 133).

When sympathetic ophthalmitis has set in we can do comparatively little. **A.** *The exciting eye,* if quite blind or so seriously damaged as to be certainly for practical purposes useless, is to be excised at once, though the evidence of benefit from this course is slender. But if it retain any moderate amount of sight it is to be left and carefully treated, since it may in the end be the more useful of the two (p. 117). **B.** *The sympathising eye.* The important measures are (1) atropine, used very often as for acute iritis; (2) absolute rest and exclusion of light by residence in a dark room and with a black bandage over the eyes; (3) repeated leeching if the symptoms are severe, or counter-irritation by blisters or by a seton in chronic cases. (4.) Mercury is believed by some to be beneficial. Quinine is sometimes given. (5.) No operation is permissible till the disease has come to a standstill; iridectomy, whilst there are active symptoms, is followed by closure of the gap with fresh lymph. The iris in bad cases often becomes very tough and adherent by its whole surface to the lens, the lens itself becoming opaque; in such cases removal of the lens and a large piece of iris by a special operation will finally be proper if the state of the eye in other respects makes it worth while.

The *prognosis* is, as will be gathered, very grave; even in the mildest cases when seen quite early we must be very cautious, for the disease often slowly progresses for many months.

Sympathetic irritation, on the other hand, is always, and as a rule promptly, cured by removal of the exciting eye; and, even if the symptoms do not subside immediately, no apprehension need be felt.

CHAPTER XI

INJURIES

INJURIES may be divided into those which affect
the eyeball itself and those limited to the surrounding
orbital structures. In each division a broad distinction is to be made between contusion and concussion
injuries, and wounds.

A. INJURIES OF PARTS AROUND THE EYEBALL.

(1.) *Contusion and concussion injuries.* — *Ecchymosis* of the skin of the eyelids from direct blows
("black eye") is to be distinguished from extravasation into the orbital cellular tissue following fracture
of the roof or other part of the orbit. In ordinary
"black eye" the ecchymosis is superficial, and, if
it affect either the palpebral or ocular conjunctiva,
does not pass far back. On the other hand, the
orbital ecchymosis following fracture is deep-seated,
often entirely beneath, rather than in, the skin and
conjunctiva, and diminishes in density towards the
front and borders of the lids; when considerable
it may cause proptosis. The two forms may be
combined when fracture is caused by direct violence
to the orbit. Cold or iced compresses, or bathing,
or an evaporating lotion will hasten absorption, if
treatment be wished for in ordinary "black eye."

Blows about the inner canthus, causing fracture of
the inner wall of the orbit into the nose, are often
followed by *emphysema of the orbital cellular tissue.*
It is to be presumed that this can occur only when
the nasal mucous membrane is perforated by the
bone. The emphysema comes on quickly from

"blowing the nose," and is shown by soft, whitish, doughy swelling of the lids, crepitating finely under the finger; the globe is more or less protruded, and its movements limited. It disappears in a few days if the lids be kept rather firmly bandaged.

Partial ptosis is an occasional result of blows upon the upper lid. It is generally accompanied by paralysis of accommodation and partial dilatation of the pupil, and it seldom lasts more than a few weeks.

But by far the most serious of the consequences which occasionally follow severe blows on the eye or the boundaries of the orbit, either immediately or after a considerable interval, are acute and chronic *orbital abscess* and *cellulitis*. Diffused acute inflammation of the cellular tissue is difficult to distinguish from acute orbital abscess, since in both there are the signs of deep inflammation, with displacement of the eye and limitation of its movements, chemosis of the conjunctiva, and brawny swelling and redness of the lids. An abscess will soon point towards some part of the eyelids, but even in cellulitis the swelling may be greater at some one part, and a feeling deceptively like fluctuation may be present.

Orbital abscess may be very chronic and simulate a solid tumour until the pus nears the surface ; even then we may be dealing with some form of cystic tumour, until an exploratory incision sets the question at rest (compare p. 51). Abscess of the orbit, whether acute or chronic, is very often the result of injury which has given rise to periostitis, and a large surface of bone is often laid bare.

In acute cases an exploratory incision is to be made with a narrow straight knife, generally through the skin (but, if practicable, through the conjunctiva) as soon as fluctuation is detected. As the pus is often curdy, it is best not to use a grooved needle. In chronic cases of doubtful nature we may wait a little. It may be necessary to go deeply into the

orbit either with the knife, probe, or dressing forceps, before matter is reached. A drainage tube
should be inserted if the abscess be deep.

The proptosis does not always disappear when an
orbital abscess is opened, because there may be much
thickening of the tissues. Sight may be injured or
lost by stretching of, or pressure on, the optic nerve,
and the cornea may become anæsthetic and may
ulcerate from damage to the ciliary nerves, behind
the globe.

(2.) *Incised and penetrating wounds.*—Wounds of
the *eyelids* need no special treatment, beyond very
careful apposition by sutures, often aided by a small
harelip pin, so as to secure primary union. Lacerated wounds of the ocular conjunctiva need a few
fine sutures if extensive, and they seldom lead to
any deformity.

Occasionally one of the recti tendons is divided or
torn through, and it will then be proper to attempt
its reunion by sutures.

Penetrating wounds through the lids or conjunctiva,
which pass deeply into the orbit, may be much more
serious than they appear at first sight, since the
stab, or thrust, or fall on to the wounding body
may have caused fracture of the orbit, and damage
to the brain membranes, or a piece of the wounding instrument may have been broken off and lie
imbedded in the roomy cavity of the orbit without
at first exciting disturbance or causing displacement of the eye. Some most extraordinary cases
are on record in which very large fragments of
iron or other substances have lain in the orbit
for a long time undetected. Again, the optic nerve
may be torn or cut across without damage to the
globe. Every wound of the eyelids or conjunctiva should therefore be carefully explored with
the probe, and whenever possible, the instrument
which caused the wound should be examined.
When a foreign body is suspected, or known to be

firmly imbedded, and is not removable through the
original wound, it is generally best to divide the
outer canthus, and prolong the incision into the conjunctiva at or about the fornix, of the upper or
lower lid according to circumstances, rather than to
divide the lid itself. In other cases an incision
through the skin, over the margin of the orbit, at
the situation of the foreign body, will be preferable.

Single shot corns, imbedded and causing no
symptoms, should not be interfered with unless they
can be easily reached.

Wounds of the orbit, by gunshot or other explo-
sions, when extensive and caused by numerous shots
or fragments of sand, gravel, &c., driven into the
tissues are serious, because the eyeball itself is often
injured, and also because tetanus may occur.

B. INJURIES OF THE EYEBALL

(1.) *Contusion and concussion injuries*

Rupture of the eye is commonly the result of severe
direct blows. The rent is nearly always in the sclerotic, a little behind or else close to the corneal
margin, with which it is concentric; the cornea
itself is but seldom rent by a blow. The rupture is
usually large, involves all the tunics, and is followed
by hæmorrhage between the retina and choroid,
and into the vitreous and anterior chambers, and
often by escape of the lens and of some of the
vitreous; sight is usually reduced to perception of
light or of large objects. The lens may escape through
a rent in the sclerotic, but be held down by the conjunctiva, if this be intact, and the rupture will then
be hidden by a prominent, rounded swelling covered
by conjunctiva. The diagnosis of ruptured eyeball
is generally easy, even if the wound be more or less
concealed. Shrinking of the eyeball, after more or
less inflammation, is a common result, but occasionally some vision is restored. Immediate ex-

cision is often best, but when there is room for hope
we should always wait until the absorption of the
blood in the anterior chamber allows the deeper parts
to be seen. The treatment will be the same as for
wounds of the eye (p. 134).

It may here be mentioned that copious hæmorrhage accompanied by severe pain sometimes occurs
between the choroid and sclerotic as the result of
sudden diminution of tension, either by an operation,
such as extraction of cataract or iridectomy, or by a
glancing wound of the cornea. Eyes in which this
occurs are for the most part already unsound and
often glaucomatous.

Blows often cause internal damage without any
rupture of the hard coats of the eye. The iris may
be torn from its ciliary attachment (*coredialysis*), so
that two pupils are formed (Fig. 26) or the lens

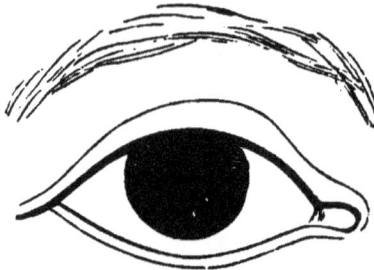

FIG. 26.—After Lawson.

be loosened or displaced by partial rupture of its
suspensory ligament, so that the iris having lost
its support will shake about with every movement
(*tremulous iris*). Such results are likely to be
attended with more or less bleeding into the anterior
chamber and into the vitreous, and the real condition
may be thus obscured for a time. The lens often
becomes opaque afterwards. Detachment of the
retina is often found after severe blows, followed by
hæmorrhage into the vitreous.

Rupture of the choroid may occur by *contrecoup*;
or, short of rupture, hæmorrhage from choroidal or

retinal vessels. These changes are found at the central part of the fundus, and very often almost exactly at the yellow spot, thus causing much damage to sight. The rents in the choroid appear, after the blood has cleared up, as lines or narrow bands of atrophy bordered by pigment, and often slightly curved towards the disc (p. 159). Hæmorrhages from the choroid into or beneath the retina, and without rupture, usually leave some pigment behind after absorption.

Paralysis of the iris and ciliary muscle, with partial and often irregular dilatation of the pupil, are sometimes the sole results of a blow on the eye. The defect of sight can be remedied by a convex lens. When uncomplicated these symptoms are seldom permanent.

Great defect of sight following a blow, and neither remedied by glasses nor accounted for by blood in the anterior chamber, will generally mean copious hæmorrhage into the vitreous, with or without the other results above mentioned in the retina and choroid. The red blood may sometimes be seen by focal light, but often its presence can only be inferred from the opaque state of the vitreous. Probably in most of these cases the blood comes from the large veins of the ciliary body, but sometimes from the choroid or from the retina. The eyeball may or may not show external ecchymosis. The tension of the globe is to be noted; it is not often increased unless inflammation has set in or the eye was glaucomatous already, and in some cases it will be below par. The prognosis should be very guarded whenever there is reason to think, from the opaque state of the vitreous, that much bleeeding has taken place, or when the iris is tremulous or partly detached, or if any rupture of the choroid can be made out. Blood in the anterior chamber is generally absorbed within about ten days, but in the vitreous this process is more tardy and less complete, permanent opacities often being left. The use of atropine, the frequent application of iced water or of an evaporating lotion to the

lids, and occasional leeching if there are inflammatory
symptoms, will do all that is possible in the early
periods.   If the lens be loosened it is likely in time
to become opaque, and it may at any time act as an
irritating foreign body, and set up a glaucomatous
inflammation, or cause sympathetic symptoms in
the other eye.   Now and then optic neuritis occurs
in the injured eye as the immediate effect of the blow.
Hæmorrhage behind the choroid may account for
some cases in which, after a blow, there is defect
of sight without visible change; and it is thought
that the mere concussion of the retina may explain
some cases of permanent defect without visible
change.

(2.) *Wounds*.—Surface scratches (*abrasions*) of the
cornea cause much pain, watering and photophobia
with ciliary congestion.   They are often caused by
a scratch by the finger nail of a baby in nursing.
The abraded surface is often very small and shows
no opacity; it is detected by watching the reflexion
of a window from the cornea (p. 5), whilst the
patient slowly moves his eye.   Now and then the eye
remains irritable for long, or becomes again troublesome after an interval.

Minute fragments of metal or stone flying from
tools, &c., may partly imbed themselves in the cornea
(*foreign body on the cornea*), and give rise to varying
degrees of irritability and pain.   If not removed
such a fragment is soon surrounded by a hazy zone
of infiltration.   Foreign bodies are easily seen unless
either very small or covered up by mucus or epithelium.   Examination by focal light (p. 25) will
show the dark speck in a doubtful case, even when
it is invisible by daylight.

The pupil is often rather smaller than its fellow,
and the colour of the iris a little altered, in cases of
abrasion and of foreign body on the cornea, these
changes indicating congestion of the iris (p. 15).

Actual iritis sometimes occurs, but not unless the
corneal wound inflames and becomes infiltrated.

Treatment.—(For removal of foreign bodies, *see*
Operations.) After surface injuries use a drop of
castor oil to lubricate the cornea, and apply a pad of
wadding and a single length of bandage tied behind
the head. Atropine is required if there is much irritation or threatened iritis. If iritis with hypopyon
arise the case will become one of hypopyon ulcer
(pp. 79 and 84).

Foreign bodies sometimes adhere to the inner surface of the lids and scratch the cornea, and the lids
must be everted, and examined whenever a patient
with a corneal abrasion states that he has " something in his eye."

Large bodies sometimes pass far back to the
upper or lower conjunctival sulcus and lie undiscovered for weeks or months, causing only local
inflammation and some thickening of the conjunctiva.
The fact is to be borne in mind and search made if
needful with a wire loop or blunt probe whenever
the suspicion arises (compare p. 124).

3. *Burns, scalds, and injuries by caustics, &c.*—The
conjunctiva and cornea are often damaged by strong
alkalies or acids, and by splashes of molten metal.
Lime, either quick or freshly slaked, is one of the
commonest agents. The eyeball is not often scalded,
the lids closing quickly enough to prevent entrance
of the steam or hot water. In none of these cases is
the full effect apparent for some days, and a cautious
opinion should, therefore, always be given when the
case is seen very early.

The effects of such accidents are manifested by
(1) inflammation, with or without ulceration, of the
cornea ; (2) scarring and shortening of the conjunctiva, and in bad cases adhesion of its palpebral and
ocular surfaces—*symblepharon ;* (3) hypopyon and
deep-seated inflammation (iritis and cyclitis) in severe
cases.

The most superficial burns whiten and dry the conjunctival surface, and in a few hours the epithelium is shed. When this desquamation affects the cornea a sharply outlined slightly depressed area is the result The floor of this denuded space is clear if the damage be quite superficial and recent, but more or less opalescent, or even yellowish, if the case be a few days old and the burn be deep enough to cause destruction or inflammation of the true corneal tissue. When there is much opacity of cornea it does not completely clear, and considerable flattening of the cornea and neighbouring sclerotic often occurs at the seat of deep and extensive burns. The conjunctival whitening is followed by mere desquamation and vascular reaction, or by ulceration and scarring, according to the depth of the damage.

Treatment.—In recent cases seen before reaction has begun a drop of castor oil once or twice a day, a few leeches to the temple, and the use of a cold evaporating lotion, or of iced water, will sometimes prevent inflammation. If seen immediately after the accident, the conjunctival sac is to be carefully searched for fragments of whatever solid has caused the mischief, or washed with very weak acid or alkaline solution if a caustic of the opposite character have done the damage. If inflammatory reaction is already present when the case comes to notice, treatment by compress, atropine, and hot fomentations, as recommended for hypopyon ulcers (p. 85), is most suitable. There is often much pain and chemosis. Prominent buttons of granulation sometimes form on the floor of a healing burn, and should be snipped off.

4. *Penetrating wounds and gun-shot injuries.*—When a patient says that his eye is wounded, the first point is to make out all we can about the wounding body, and especially whether or not any fragment has been left within the eyeball. We next examine carefully the seat, extent and character of the wound, ascertain

the interval since the injury, and test the sight of the eye.

Treatment.—Penetrating wounds are least serious when they implicate the cornea alone, or the sclerotic alone behind the ciliary region, *i.e.* when situated at least $\frac{1}{6}$ inch behind the cornea. Penetrating wounds of the cornea, without injury to the iris or lens, and without any prolapse of iris, are rare ; they generally do very well, and if the case is not seen until two or three days after the injury, the wound will often have healed firmly enough to retain the aqueous, and it may be difficult to decide whether the whole thickness of the cornea was penetrated or not. Wounds of the sclerotic seldom unite without the interposition of a layer of lymph ; if seen early they may sometimes, if there is reason to think that the eyeball is worth preserving, be advantageously treated by the introduction of one or two fine sutures, followed by constant application of ice.

But penetrating wounds usually are very serious to the injured eye : the iris is frequently lacerated and included in the track of the wound ; the lens is punctured and becomes swollen and opaque from absorption of the aqueous tumour (*traumatic cataract*), and liable in its swollen state to press on the ciliary processes and cause grave symptoms ; extensive bleeding perhaps takes place into the vitreous ; at a later period, plastic or purulent cyclitis may destroy the eye. The fellow eye is of course often in danger of sympathetic inflammation (p. 115). Every case has therefore to be judged from two points of view, the damage to the injured eye and the risk to the sound one, and the question of sacrificing or attempting to save the former is sometimes very difficult to decide.

Very large foreign bodies, such as pieces of glass, sometimes lie for long in the eye without causing much trouble ; the large wound having given exit

to the contents of the globe and been followed by
rapid shrinking without inflammation.

(A.) In the two following cases the eye should be
excised at once. (1.) If the wound, lying wholly
or partly in the "dangerous region" (p. 116), is so
large and so complicated with injury to deeper parts
that no hope of useful sight remains. (2.) If, even
though the wound is small, it lies in the dangerous
region, and have already set up cyclitis. Cyclitis
will be shown by intense ciliary congestion, tenderness and pain, very great defect of sight, and (if the
lens be transparent) a yellow or greenish reflex from
the vitreous.

(B.) There is a large class of cases in which the
wound, though in the ciliary region, or involving the
lens and iris through the cornea, is not of itself fatal
to sight and has not as yet led to inflammation or
to shrinking of the eye.

The first question will then be whether or not the
eye contains a foreign body, and the second whether
the lens is wounded. A foreign body, if lying on or
embedded in the iris, the lens being intact, should be
removed, usually with the portion of iris to which it
is attached. If it can be seen in the lens and the
condition of the eye be otherwise favorable, a scoop
extraction may be done in the hope of removing the
fragments with the lens; in a few cases a powerful
magnet has been employed for the extraction of chips
of iron from the lens and iris, and may be occasionally
useful in eyes otherwise worth saving. If it is certain that the foreign body has passed into the vitreous, whether through the lens or not, and whether
by gun-shot or not, it is scarcely ever worth while
to attempt to save the eye. The body can of course
seldom be seen, but a track of opacity through the
lens with extensive hæmorrhage into the vitreous,
or even the latter alone, with conclusive history
that the wound was made by a fragment or a
shot, and not by an instrument or large body, is

generally enough to settle the point in favour of excision.

(c.) There remain cases of wound without retention of foreign body in the eye ; (1) the wound being in the dangerous region and complicated with traumatic cataract ; (2) in the dangerous region without traumatic cataract ; (3) with traumatic cataract, but the wound being corneal, and, therefore, out of the dangerous zone. In the first, and still more in the second of these, there will often be much difficulty in deciding what to do, it being presumed that the wounded eye shows no signs of severe internal inflammation (cyclitis). Some of the most difficult cases are those of group (2) of wounds by sharp instruments close to the corneal border, with considerable adhesion of the iris, or in which there is evidence that the track lies between the lens and the ciliary processes, the lens not being wounded, and useful sight remaining. If the patient is seen within two or three weeks of the injury, and the sound eye shows no irritation, we may safely watch the case for a few days. If decided sympathetic irritation (see p. 117) be present and do not yield after a few days' treatment, excision is advisable, even though the lens of the wounded eye is uninjured. I think that if we made a rule of excising every eye with wound in the ciliary region and traumatic cataract, whether or not it were causing sympathetic symptoms or were itself especially irritable, we should not be far wrong, for the prospect of regaining useful vision in the eye under such circumstances is but slight. In the third group, excision is justifiable only if severe iritis and cyclitis come on with threatened panophthalmitis, and this seldom occurs. The patient in all open cases must be warned, and must be seen every few days for many weeks.

When sympathetic ophthalmitis (p. 117) has set in before the patient asks advice, the rule for excision of the exciting (wounded) eye is different.

It is then to be removed only if there is no reason to hope for restoration of useful sight in it; a moderate degree of subacute irido-cyclitis with or without traumatic cataract, and with sight proportionate to the state of the lens, is no indication for excision, since this eye may very probably in the end be the better of the two (see p. 116).

The treatment of wounded eyes which are not excised is the same as for traumatic iritis, viz., atropine, rest, darkness, and local depletion and counter-irritation (compare p. 120). At the commencement the constant use of ice is a very important means for lessening the severity of the symptoms.

It is sometimes important to determine whether an eye which has been excised contains a foreign body. If nothing can be found in the blood or lymph, &c., by feeling with a probe or needle, it is best to crush up the soft parts carefully, little by little between finger and thumb, when the smallest particle is almost sure to be felt. If a shot has entered and left the eye, the counter-opening may, if recent, be found by a probe from the inside of the globe although no irregularity be noticeable on the outer surface of the sclerotic.

CHAPTER XII

CATARACT

CATARACT means opacity of the crystalline lens and is due to changes in the structure and composition of the lens fibres, the capsule not being materially altered. These changes seldom occur throughout the whole lens at once, but commence first in some one part, *e. g.* the centre (nucleus) or the superficial layers (cortex), whilst in some of the forms of partial cataract the disease remains permanently confined to some well-circumscribed part.

Senile changes in the lens.—With advancing age the lens, which is from birth firmest at the centre, becomes harder and flatter, its refractive power changes and its surface reflects more light, the nucleus acquires a yellowish colour, and the substance of the lens becomes somewhat fluorescent. The result of all these changes is that at an advanced age the lens is often more easily visible than in early life, the pupil becoming greyish instead of quite black. This greyness of the pupil may easily be mistaken for cataract, but as it is compatible with almost the normal transparency, the fundus in such a case being seen without any appreciable blurring; such a mistake in diagnosis ought never to be made.

The consistence of a cataract depends more on the patient's age than on the position or character of the opacity. Below about thirty-five all cataracts are "soft." When more is known of the causes of generalised opacity of the lens we shall probably find that these are often the same at different times of life, and that the wide physical differences between cataracts depend less on variations in the cause than on the degree of natural hardness the lens possesses when the opacity sets in.

FORMS OF GENERAL CATARACT

(1.) *Nuclear cataract.*—The opacity begins in and remains more dense at the centre (nucleus) of the lens, thinning off gradually in all directions towards the cortex (Fig. 29) ; the nucleus is not really opaque, but densely hazy like thick fog. The patients are generally old people in whom the nucleus is naturally very firm and yellowish ; hence nuclear cataract is also usually senile and hard, to which we may add that it is often amber-coloured or light brownish, or like a " peasoup " fog.

(2.) *Cortical cataract.*—The change begins in the superficial parts, and generally in the form of sharply defined lines or streaks, or triangular patches which point towards the axis of the lens, and whose shape is dependent on the arrangement of the lens fibres (Fig. 30). They usually begin at the edge (equator) of the lens where they are hidden by the iris, but when large enough encroach on the pupil as whitish streaks or triangular patches. They affect both anterior and posterior surface of the lens, and the intervening parts may be quite clear. Sooner or later the nucleus also becomes hazy (mixed cataract), and the whole lens eventually gets opaque.

Almost all the large class known as " senile " or " hard " cataract are either nuclear from beginning to end, *i. e.* formed by gradual extension of diffused opacity from the centre to the surface, or else of the mixed variety.

A few cataracts beginning at the nucleus, and many beginning at the cortex, are not senile in the sense of accompanying old age, and are, therefore, not hard. Some such are caused by diabetes, but in many it is impossible to say, excepting by a general reference to bad health or premature senility, why the lens should have become diseased. Many such are known as "soft" cataracts when complete. They generally form quickly in a few months. A few are congenital. Whether nuclear or cortical, they are whiter and more uniform looking than the slower cataracts of old age, and the cortex often shows a linear glisten like satin, or a flaky appearance like crystallised spermaceti.

In some cortical cataracts we find only an immense number of very small dots or short streaks (dotted cortical cataract). Occasionally a single large wedge-shaped opacity will form at some part of the cortex and remain stationary and solitary for many years. Sometimes, though no opaque striæ are visible by focal illumination, one or more dark streaks are seen with the mirror which alter as it is differently inclined. These "flaws" in the lens are comparable in their optical effect to cracks in glass, and must always be looked on as the beginning of cataract.

PARTIAL CATARACTS

Three forms need special notice.

(1.) *Lamellar (zonular) cataract* is a peculiar and well-marked form in which the superficial laminæ and nucleus of the lens are clear, a layer or shell of opacity being formed between them (Fig. 32). It is probably formed during the first few months of life, or it may in some cases be congenital. The great majority of its subjects suffer from infantile convulsions. The size of the opaque lamella or shell, and, therefore, its depth from the surface of the lens, is subject to much variation, and it may be much smaller than is shown in Fig. 32. The opacity is often stationary for many years, sometimes, perhaps, for life ; in other cases the little spicules which are often seen projecting towards the cortex increase in number and size, and the cataract eventually becomes general.

(2.) *Pyramidal cataract.*—A small, sharply-defined spot of chalky-white opacity is present in the middle of the pupil, looking as if it lay upon the capsule. When viewed sideways it seems to be superficially embedded in the lens, and also sometimes stands

forwards as a little nipple or pyramid (Fig. 27). It consists of the degenerated products of a localised inflammation just beneath the lens capsule, with the addition of organised lymph derived from the iris and deposited on the front of the capsule, the capsule

FIG. 28.—Magnified section of a pyramidal cataract. The parallel shading represents the thickness of the opacity, the double (black and white) outline is the capsule ; on each side are the cortical lens fibres, many being broken up into globules beneath the opacity. Lying upon the puckered capsule over the opacity is a little fibrous tissue, the result of iritis.

itself being puckered and folded (Fig. 28). It is always stationary, and never becomes general.

Pyramidal cataract is the result of central perforating ulceration of the cornea in early life, and ophthalmia neonatorum is nearly always the cause of the ulceration. It is generally associated with central opacity of the cornea. The contact between the exposed part of the lens capsule and the inflamed cornea, which occurs when the aqueous has escaped through the hole in the ulcer, appears to set up the localised subcapsular inflammation. It has been suggested that the same change may occur without ulceration by the transmission of influence from the inflamed conjunctiva through the non-ulcerated cornea* ; and it is probable that iritis in early life may cause similar subcapsular spots. The term anterior polar cataract is also applied to the pyramidal form and to some less common varieties which begin in the same part of the lens.

(3.) Cataract which afterwards becomes general may begin at the middle of the hinder surface of the lens (*posterior polar cataract*) (Fig. 31). There are many varieties, but in general the pole itself shows the most change, the opacity radiating outwards from it in more or less regular spokes, and appearing quite thin, a mere lamina. The colour is greyish, yellowish, or even brown, as seen through the whole thickness of the lens. Sometimes the opacity is situated really just behind the capsule, *i. e.* in the hyaloid membrane or front of the vitreous, but this cannot be proved during life.

Cataract beginning at the posterior pole is often a sign of disease of the vitreous depending on choroidal disease ; it is common in the later stages of retinitis pigmentosa and of severe choroiditis, and in high degrees of myopia with disease of the vitreous. The prognosis, therefore, should always be guarded in a case of cataract where the principal part of the opacity is in this position.

When a cataract forms without known connection with other disease of the eye it is called primary. The term *secondary cataract* is used when it is the consequence of some local disease, such as severe iridocyclitis, glaucoma, detachment of the retina, or the growth of a tumour in the eye. The pyramidal cataract is strictly a secondary form, though not usually spoken of as such. Primary cataract is almost always symmetrical, though seldom syn-

* Hutchinson.

chronous in the two eyes; whilst secondary cataract, of course, may or may not be symmetrical.

The subjective symptoms of cataract depend solely on the obstruction and distortion of the entering light by the opacities. Objectively cataract is shown in advanced cases by the white or grey condition of the pupil at the plane of the iris; in earlier stages by whitish opacity in the lens when examined by focal illumination (p. 24) and by corresponding dark portions (lines, spots or patches) in the red pupil when examined by the ophthalmoscope mirror.

Both subjective and objective symptoms differ with the position and quantity of the opacity. When the whole lens is opaque the pupil is uniformly whitish; the opacity lies almost on a level with the iris, no space intervening, and, consequently, on examining by focal light we find that the iris casts no shadow on the opacity; the brightest light from the mirror will not penetrate the lens in quantity enough to illuminate the choroid, and hence no red reflex will be obtained. Such a cataract is said to be mature or "ripe," and the affected eye will be in ordinary speech "blind." If both are equally affected, the patient will be unable to see any objects; but he will distinguish quite easily between light and shade when the eye is alternately covered and uncovered in ordinary daylight (Good perception of light, *p. l.*), and will see the position of a candle flame.

Diagnosis of immature and partial cataracts

The patient complains of gradual failure of sight, and we find the acuteness of vision (p. 18) impaired more or less (probably more in one eye than in the other). If he can still see moderate types the glasses appropriate for his age and refraction, though giving some help, do not remove the defect, whilst for distant objects vision is likely to be worse in proportion than for the near types. If, as is usual, he be presbyopic, he will be likely to choose over-strong spectacles, and to place objects too close to his eyes, so as to obtain larger retinal images, and thus compensate for want of clearness.

In nuclear cataract, as the central (axial) rays of light are most obstructed, sight is often better when the pupil is rather large, and such patients tell us that they see better in a dull light or with their back to the window, or when shading the eyes with the hand. In the cortical and more diffused forms this symptom is less marked or quite wanting.

On examining by focal light (after dilating the pupil with atropine) an *immature nuclear cataract* will appear as a yellowish, rather deeply-seated haze, on which a shadow will be cast by the iris on the side

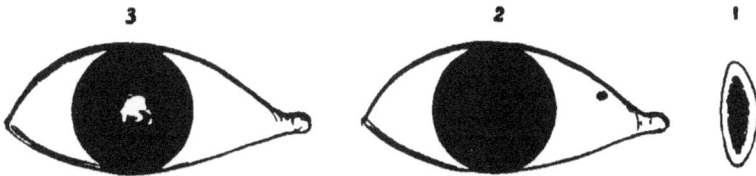

FIG. 29.—Nuclear cataract. 1. Section of lens; opacity densest at centre. 2. Opacity seen by transmitted light (ophthalmoscope mirror) with dilated pupil. 3. Opacity as seen by reflected light (focal illumination).

from which the light comes (3, Fig. 29). On now using the mirror this same opacity will appear as a dull blur in the area of the red pupil, darkest at the centre, and gradually thinning off on all sides, so that, at the margin of the pupil, the red choroidal reflex will be quite bright (2, Fig. 29); when the opacity is very dense and large only a faint dull redness will be visible quite at the border of the pupil. If the fundus can be seen it will be as through a fog, but a fog thickest in the axis of vision, so that by looking through the more lateral parts the details will be better seen.

Cortical opacities, if small and confined to the equator (or edge) of the lens, do not interfere with sight; they are easily detected with a dilated pupil

by throwing light very obliquely behind the iris.
When large and encroaching on the pupil they are
visible in ordinary daylight. They are in the form
of dots, streaks, or bars; seen by focal light they are
white or greyish, and appear more or less sharply
defined, according as they are in the anterior or
posterior layers (3, Fig. 30). With the mirror they

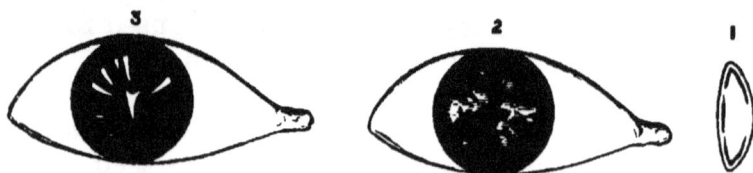

FIG. 30.—Cortical cataract.   References as in preceding figure.

appear black or greyish, and of rather smaller size
(2, Fig. 30), and if the intervening substance is clear,
the details of the fundus can be seen sharply between
the bars of opacity.

Posterior polar opacities are seldom visible without
careful focal illumination, when we find a patchy or
stellate figure very deeply seated in the axis of the
lens (3, Fig. 31); if large it will look concave
like the bottom of a shallow cup. With the
mirror it will be seen as a dark star (2, Fig. 31),

FIG. 31.—Posterior polar cataract.   References as before.

or network, or irregular patch, rather smaller than
when seen by focal light.

The diagnosis of *lamellar cataract* is easy when its
nature is understood, but otherwise it may be readily
diagnosed as "nuclear." The patients are generally
children or young adults; they complain of "near

sight" rather than of "cataract" or "blindness;" for the opacity is not usually very dense, and whether the refraction of their eyes be really myopic or not (and it often is so), they (like other cataractous patients) compensate for dull retinal images by holding the object nearer, and so increasing the size of the images. The acuteness of vision is always defective, and cannot be fully remedied by any glasses. Varying brightness of the light affects their sight but little, for not only is the cataract uniformly dense in all parts, but its diameter is usually greater than that of the pupil, even in dull light, and they, therefore, can make no use of the clear margin. The pupil presents a deeply-seated slight greyness (4, Fig. 32), and when dilated with atropine the outline

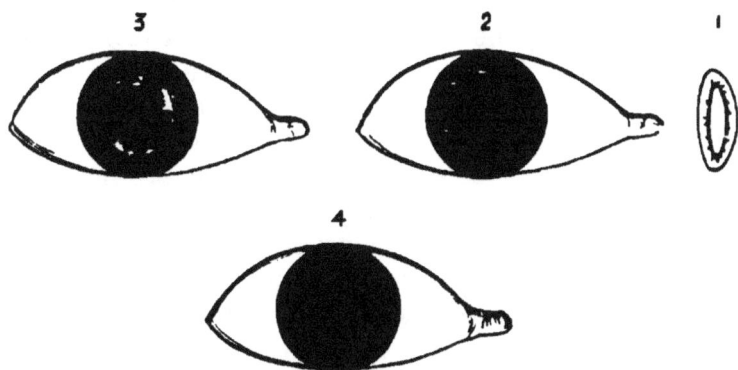

FIG. 32.—Lamellar cataract. Figs. 1, 2, 3, as before. Fig. 4 shows *slight* greyness of the undilated pupil owing to the layers of opacity being deeply seated.

of the shell of opacity is exposed within it. It is sharply defined and circular, and by focal light is whitish, interspersed, in many cases, with white specks, which at its equator appear as little projections (3, Fig. 32). By this examination we easily make out that the opacity consists of two distinct layers, that there is a layer of clear lens substance (cortex) in front of the anterior layer, and that the margin (equator) of the lens is clear. By the mirror the opacity appears as a disc of nearly uniform greyish or dark colour, with or without projections, and darker dots, and surrounded by a zone of bright red reflexion from the fundus corresponding to the clear margin of the lens (2, Fig. 32). The opacity often appears rather more dense just at its boundary, a sort of ring being formed there.

In some cases quite large spicules or patches project from the margin of the opacity. Not only does the size of the opaque lamella, and, therefore, its depth from the surface of the lens, differ greatly in different cases, but its thickness or degree of opacity varies also. The disease is nearly always exactly symmetrical in the two eyes. Occasionally there are two shells of opacity, one within the other, separated by a certain amount of clear lens substance.

The lens may be cataractous at birth (*congenital cataract*). This form, of which there are many varieties, is nearly always symmetrical, and generally involves the whole lens. Often the development of the eyeball is defective, and though there are no synechiæ, the iris often acts badly to atropine.

Severe blows on the eye may be followed by opacity of the lens, the suspensory ligament being generally torn in some part of its circle (*concussion cataract*), but I am not aware that cataract ever follows injury to the head without direct injury to the eye.

Traumatic cataract proper is the result of wound of the lens capsule ; the aqueous passing through the aperture is imbibed by the lens fibres, which swell up, become opaque, and finally disintegrate and are absorbed. The opacity begins within two or three days of the wound and progresses quickly in proportion as the wound is large, and the patient young. The older the patient the more severe are the accompanying symptoms likely to be, and the worse the prognosis. A free wound of the capsule followed by rapid swelling and opacity of the whole lens, in an adult past middle life, may give rise to severe glaucomatous symptoms and iritis. In from three to about six months the wounded lens will generally be absorbed and nothing but some chalky white detritus remain in connection with the capsule. A very fine puncture of the lens is occasionally followed by nothing more than a small patch or narrow track of opacity, or by very slowly advancing general haze.

The objects of *treatment* are to prevent iritis and posterior synechiæ by atropine, and by ice and leeching if there be severe inflammatory symptoms. We endeavour to wait for the natural absorption of the cataract, being prepared to extract the lens with iridectomy or by suction, at any time, should persistent glaucomatous symptoms arise.

Prognosis in cataract.—Cataracts advance with very varying rapidity in different cases. As a rough rule, the progress of a general cataract is rapid in proportion to the youth of the patient. Cataracts in old people commonly take from one to three years in reaching maturity. If the lens be allowed to remain for long after it is opaque, further degenerative changes may occur, calcareous particles being formed, and the lens substance partially liquefied; finally, after absorption of the fluid it may shrink to a thin, hard, brittle disc. Soft cataract in young adults, whether from diabetes or not, is generally complete in a few months.

The prognosis for sight after operation is good when there is no other deep-seated disease of the eye, and when the patient (although advanced in years) is in fair general health. It is not so good in diabetes, nor when the patient is comparatively young, and in obviously bad health, because the eyes are then less tolerant of operative interference. In lamellar and congenital cases it must be guarded, for the eyes are often defective in other respects, and sometimes very intolerant of operation; the intellect,

10

too, is sometimes defective, rendering the patient less able to make proper use of his eyes.

In traumatic cataract of course everything depends on the details of the injury (*see* Injuries, p. 130, &c.), but in general the younger the patient the better the prospect of a quiet and uncomplicated absorption of the lens.

In every case of immature cataract, the vitreous and fundus should be carefully examined by the ophthalmoscope, and the refraction ascertained. The presence of high myopia is unfavorable, and the same is true of opacities in the vitreous, indicating as they usually do that it is fluid. Any disease of the choroid or retina will, of course, act injuriously in proportion to its position and degree. In every case, whether complete or not, the size and mobility of the pupils to light and atropine and the tension of the eye are to be carefully noted.

Treatment.—In the early stages of nuclear cataract sight is improved by keeping the pupil moderately dilated with a weak solution of sulphate of atropine (half a grain to the ounce), used about three times a week. Dark glasses, by allowing some dilatation of the pupil, sometimes give a little relief in the same way. With these exceptions, nothing except operative treatment is of any use.

The management of lamellar cataract requires separate description.

Operations for the removal of cataract are of three kinds: (1) Extraction of the lens entire through a large wound in the cornea, or at the sclero-corneal junction, the lens-capsule remaining behind. By a few operators (Pagenstecher and Macnamara), the lens is removed entire in its capsule. (2) Gradual absorption of soft cataracts by the action of the aqueous humour, admitted through needle punctures in the capsule, just as after accidental traumatic cataract (needle operations, solution, discission). The operation needs repetition two or three times, at in-

tervals of a few weeks, and the whole process therefore spreads over three or four months. (3) Removal
of soft cataracts *en masse,* by a suction syringe or
curette, introduced into the anterior chamber·through
a small wound near the margin of the cornea, the
whole lens having been rendered semifluid by performing a free needle operation a few days before.
(For details *see* Operations.)

Extraction is necessary for cataracts after about
the age of forty, the lens from this age onwards being
so firm that its absorption by needle operations takes
a much longer time than in childhood and youth ;
moreover, the swelling of the lens after puncture by
the needle is less easily borne as age advances, and
solution operations therefore become not only slower
but attended by more danger (*see* Traumatic Cataract). Indeed, extraction is often practised in preference to solution some years earlier than forty. Suction
and solution operations are applicable up to about
the age of thirty-five.

The suction operation is difficult, and unless we
performed is attended by much risk of severe iritis
and cyclitis. Its advantage is the saving of time,
as compared with needle operations, the whole lens
being removed at one sitting.

So long as senile cataract is single, or if double, so
long as the second eye is still serviceable, removal of
the cataract will seldom be beneficial to the patient
unless his health be likely to suffer by waiting till
the second eye is ready and his prospect of a good
result be thus impaired. Indeed, if one eye be still
fairly good, the patient will often be dissatisfied by
finding his operated eye less useful than he expected,
perhaps even not so useful as the other. The
removal of a single cataract in young persons is
quite justifiable on the ground of appearance alone,
and sometimes expedient when ʰrom his occupation
it is important that the patient should not have a
" blind side." But in such cases the patient must

be made to understand that after the operation the
two eyes will not work together on account of
the extreme difference of refraction (*see* Anisometropia).

Even when both cataracts are mature at the same
time it is safer to remove only one at once, because
the after treatment is more easily carried out upon
one eye than both, and because after double operation
any untoward result in one eye adds to the difficulty
of managing its fellow; while a bad result after
single extraction enables us to take especial precautions for the second eye.   Even if the patient be so
old or feeble that the second eye may never come to
operation, we shall consult his interests better by
endeavouring to give him one good eye than by
risking a bad result in attempting to give him both
at the same time.

The bad results to be feared after extraction are
chiefly—

(1.) *Sloughing of the cornea,* a very rare event,
except after flap-extraction, and setting in within
twenty-four hours of the operation.

(2.) Various degrees of *suppurative inflammation,*
beginning in the corneal wound, and in most cases
spreading to the whole cornea, to the iris and vitreous,
and ending in total loss of the eye.   It occasionally
takes a less rapid course, and stops short of a fatal
result.   The alarm is given about the second or third
day by the occurrence of pain, inflammatory œdema of
the lids (particularly the free border of the upper lid),
and the appearance of some muco-purulent discharge.
On raising the lid the eye is found to be greatly congested, and its conjunctiva œdematous, the edges of
the wound yellowish, and the neighbouring cornea
steamy and hazy.   In very rapid cases the pupil,
especially near to the wound, may already be occupied
by lymph.

The energetic use of hot fomentations for an hour

three or four times a day, and the contant employment
between times of a firm compressive bandage, are the
only local means likely to be useful, while internally
full doses of quinine with ammonia, and wine or
brandy, should be at once resorted to. But the great
majority go on to suppurative panophthalmitis or to
severe plastic irido-cyclitis with opacity of cornea and
shrinking of the eyeball.

(3.) *Iritis* (sometimes called " primary " because
supposed not to be propagated from the corneal
wound) may set in on the fifth day or several days
later. As in commencing suppuration, so here, pain,
œdema of the eyelids, and chemosis are the earliest
symptoms. There is lachrymation, but no muco-purulent discharge, and the cornea and wound remain clear
and bright. The iris is discoloured (unless it happen
to be naturally greenish brown), and the pupil will
not dilate fully or evenly to atropine. Whenever in
a case presenting such symptoms a good examination
is rendered difficult on account of the photophobia,
iritis should be suspected. If the early symptoms
are severe, a few leeches to the temple are very
useful. Atropine and local warmth are the most
important remedial measures. If atropine after a
time sets up its characteristic irritation, daturine
must be used.

This iritis is plastic, and ends in the formation
of more or less dense membrane which occupies the
area of the pupil, and often, by contracting and drawing
the iris with it towards the operation scar, diminishes
and displaces the pupil.

(4.) The iris may prolapse into the wound at the
operation, or a few days afterwards by the reopening
or yielding of a weakly united wound. When iridectomy has been done the prolapse appears as a little
dark bulging at the angles of the wound, and often
causes much irritability for many weeks without
actual iritis. The protrusion in the end generally
flattens down, but sometimes it needs to be punc-

tured or even removed. The occurrence of prolapse is a reason for keeping the eye tied up longer.

After operations are needed, when iritis has ended in more or less occlusion and contraction of the pupil. Nothing should be done until all active symptoms have subsided, and the eye has been quiet for some weeks. In slight cases gently tearing the thin membrane with a fine needle, and treating the case as after other needle operations, will do all that is needed, and often wonderfully improve the sight. When the membrane is thick, and has by its contraction stretched the iris fibres towards the operation scar, we must either make an iridectomy in the opposite direction, or cut the iritic membrane and adjacent iris with a specially designed scissors passed through a wound in the cornea. After the latter operation (iridotomy) a clear pupil is formed by the retraction of the cut edges of the tense iris.

Sight after the removal of cataract.—In accounting for the state of the sight we have to remember that the acuteness of sight naturally decreases in old age, being at sixty about three quarters, and at eighty little more than half of the standard taken as normal. Again, slight iritis producing a little filmy opacity in the pupil is common after extraction. Lastly, some eyes without positive inflammation remain irritable for long after the operation, so that prolonged use is impossible. So that putting aside the graver complications we find that even of the eyes which do best a large proportion fail to reach anything like normal acuteness of vision. Cases are considered good when the patient can with his glasses read anything between Nos. 1 and 14 Jaeger and $\frac{20}{70}$ Snellen; but a much less satisfactory result than this is a great gain both to the patient and his friends.

The eye is rendered extremely hypermetropic by removal of the lens, and strong convex glasses are necessary for clear vision. They should not be allowed until three months after the operation, and at first they must not be continuously worn. Two pairs are needed, one making the eye emmetropic, and giving clear vision of distant objects (about $+ 3''$ or $3\frac{3}{4}''$ focal length), the other (about $+ 2\frac{1}{2}''$ focal length) making the eye myopic, and giving clear vision of objects at a specified short distance ($8''$ or $10''$). As all accommodation is lost, the patient has scarcely any *range* of distinct vision.

Lamellar cataract.—If the patient can see enough to get on fairly well at school or in his occupation, it is best not to destroy the lenses. When the opacity is dense enough to interfere seriously with the patient's prospects, something must always be done. The choice lies between artificial pupil when the margin of clear lens is wide, and solution or extraction when it is narrow, or when large spicules of opacity project into it from the opaque lamella. It is very difficult to say which of the two gives, on the whole, the better results, and each case must be judged on its own merits based particularly on whether or not atropine, by dilating the pupil, improves the sight; if it does so an artificial pupil will generally be beneficial. A very good rule is to operate on only one eye first, thus allowing the choice of a different operation on its fellow.

Secondary cataracts with complete blindness, from deep-seated disease, should never be operated on.

CHAPTER XIII

DISEASES OF THE CHOROID

THE choroid is, next to the ciliary processes, the most vascular part of the eyeball, and from it the outer layers of the retina, and probably the vitreous humour also, mainly derive their nourishment. Inflammatory and degenerative changes often occur, some of them entirely local, as in myopia, others symptomatic of constitutional or of generalised disease, such as syphilis and tuberculosis. Disease of the choroid, unlike disease of its continuations, the ciliary body and iris, is seldom attended by congestion, heat, or pain. Its diagnosis rests almost entirely on ophthalmoscopic evidence, the subjective symptoms not being pathognomonic.

Blemishes or scars, permanent and easily seen, nearly always follow disease of the choroid, and such spots and patches are often as useful for diagnosis as cicatrices on the skin, and deserve as careful study. In certain cases, the retina lying over an inflamed choroid takes part in the changes, or becomes atrophied afterwards; whilst in others, apparently as severe, it is uninjured. Indeed it is sometimes far from easy to say in which of these two structures the disease has begun, especially as changes in the pigment epithelium, which is really part of the retina, are as often the result of deep-seated retinitis or hæmorrhage as of superficial choroiditis. Lastly, patches of accumulated pigment may represent either spots of former choroiditis or extravasations of blood into the retina, and some skill is needed in drawing correct conclusions from such changes.

Appearances in health.—The choroid is composed chiefly of blood-vessels and of cells containing dark brown pigment. The quantity of pigment varies much in different eyes, and to some degree in different parts of the same eye; it is very scanty in early childhood, and in persons of fair complexion; more abundant in persons with dark hair and brown irides; more plentiful in the region of the yellow spot than elsewhere. In old age the pigment epithelium becomes paler. When examining the choroid we need to think of four parts: (1) the retinal pigment epithelium (which is for ophthalmoscopic purposes choroidal), recognised in the erect image as a fine darkish stippling; (2) the capillary layer (chorio-capillaris), just beneath the epithelium, forming a very close meshwork, the separate vessels of which are not visible in life; (3) the larger blood-vessels often easily visible; (4) the pigmented connective-tissue cells of the choroid proper, which lie amongst the larger vessels.

In the majority of eyes these four structures are so toned as to give a pretty uniform full red colour by the ophthalmoscope, blood colour predominating. In very dark races the pigment is so excessive as to give an uniform slaty colour to the fundus of the eye; in very fair persons (and young children) the deep pigment (4) is so scanty that the large vessels are separated by spaces of lighter colour than themselves; whilst in dark individuals these intervascular spaces are of a deeper hue than the vessels, the latter appearing like light streams separated by dark islands. Near to the disc and yellow spot the vessels are extremely abundant and very tortuous, the interspaces (whether light or dark) being small and irregular; but towards and in front of the equator the veins take an approximately parallel course, converging to their exits at the *venæ vorticosæ*, and the islands (whether light or dark) are larger and elongated. The veins are much more numerous and larger than the arteries (Fig. 34), but no distinction can be made between them in life.

The pigment epithelium and the capillary layer tone down the above contrasts, and so in old age, when the pigment of the epithelium is atrophied, and again when the capillary layer is atrophied after superficial choroiditis, the distinctions described are particularly marked. Fig. 33 shows a vertical section of naturally injected human choroid : the uppermost dark line is the pigment epithelium (1) ; next are seen the capillary vessels, cut across (2) ; then the more deeply-seated large vessels (3), and the deep layer of stellate pigment-cells of the choroid proper (4). Fig. 34 is from an artificially injected human choroid seen from the inner surface. The shaded portion is intended to represent the general effect produced by all the vessels and the pigment epithelium. The lower part shows the large vessels with their elongated interspaces, as may be seen in the eye of a blonde with atrophy of the pigment epithelium and choriocapillaris ; in a dark eye these interspaces would be darker than the vessels ; the arteries are cross-shaded. The middle part shows the capillaries without the pigment epithelium ; the capillaries are engraved rather too small. Both figures are magnified twenty diameters, about four times as much as the image in the indirect ophthalmoscopic examination.

FIG. 33.

Ophthalmoscopic signs of disease of choroid.—The changes most usually met with are indicative of atrophy. This may be partial or complete, primary or following inflammation or hæmorrhage, in circumscribed spots or patches, or in large and less abruptly bounded areas ; and there are very often secondary changes in the corresponding parts of the

retina.   The chief signs of atrophy of the choroid
are—(1) the substitution of a paler colour (varying

FIG. 34.

from a pale red to a full paper white), for the full
red of health, the subjacent white sclerotic being more
or less visible where the atrophic changes have occurred ; (2) pigment in the form of spots, patches, or
rings in varying quantity upon, or around the pale
patches. These pigmentations result, 1st, from distur-

bance and heaping together of the normal pigment;
2nd, from increase in its quantity; 3rd, from colouring
matter imperfectly absorbed after extravasations of
blood. Patches of primary atrophy (*e. g.* in myopia)
are never much pigmented unless bleeding have
taken place. The amount of pigmentation in atrophy
following choroiditis is closely related to that of the
healthy choroid, *i. e.* to the complexion of the person.

Pigment in the fundus may lie in the retina as well
as in or on the choroid, and this is true whatever may
have been its origin, for in choroiditis with secondary
retinitis the choroidal pigment often passes forwards
into the retina. When a spot of pigment is distinctly seen to cover over a retinal vessel that spot
must, of course, be not only in the retina, but very
near to its anterior surface; and when the pigment
has a linear or moss-like or lace-like pattern
(Fig. 41), it is always in the retina. These are the
only conclusive evidences of its position.

It is important and usually easy to distinguish

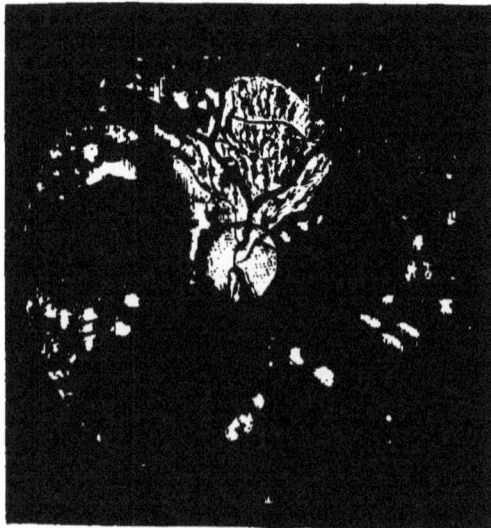

FIG. 35.—Atrophy after syphilitic choroiditis, showing various
degrees of partial atrophy at the upper part and patches of
more advanced atrophy in other parts (Hutchinson).

between partial and complete atrophy of the choroid. In superficial atrophy affecting the pigment epithelium and capillary layer, the large vessels are peculiarly distinct (Fig. 35, upper part).   Such "capillary" or "epithelial" choroiditis often covers a large surface, the boundaries of which are sometimes well defined and sinuous or map-like, but are as often ill marked.   In badly marked cases a careful comparison between different parts of the fundus, taken with the patient's age and complexion, may be needful to establish a conclusion. Complete atrophy is shown by the presence of patches of white or yellowish-white colour of all possible variations in size, with sharply cut, circular, or undulating borders, and with or without pigment accumulations (Figs. 35 and 36). The retinal vessels which run over such patches are

FIG. 36.—Atrophy after choroiditis (Magnus).

unobscured, proving that the appearance is caused by some change deeper than the surface of the retina.

In recent choroiditis we may also often see patches of palish colour, but they are less sharply bounded and frequently of a greyer, or bluer, or whiter (less yellow) colour than patches of atrophy ; their edges

are softened, the texture of the choroid being at those
parts dimly visible because only partly veiled by
exudation.　If the overlying retina is unaffected
its vessels are clearly seen on the part diseased, but
often the retina itself shows hazy or opaque patches,
and the exact seat of the exudation cannot then be
at once settled.　In recent cases the vitreous too is

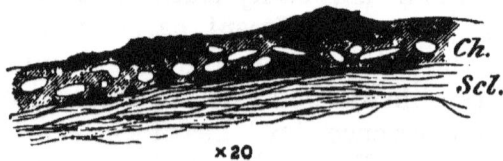

×20

FIG. 37.—Minute exudations into inner layer of choroid in
syphilitic Choroiditis.　Pigment epithelium adherent over
the exudations, but elsewhere has been washed off. *Ch.*
Choroid; *Scl.* Sclerotic.

often hazy or full of flocculi.　Patients, however,
do not often come to us until the exudation stage of
choroiditis has passed into atrophy.

Syphilitic choroiditis generally begins in, and is
often confined to, the inner (capillary) layer of the
choroid (Fig. 37), and hence it often affects the
retina.　In miliary tuberculosis of the choroid the
overlying retina is clear, and the growth is, for the
most part, deeply seated around an artery (Fig. 38).

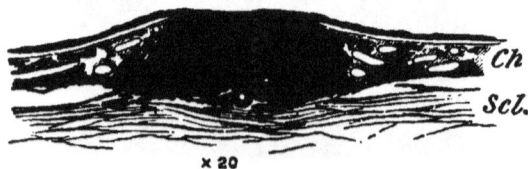

×20

FIG. 38.—Section of miliary tubercle.　Inner layers of choroid
comparatively unaffected. The lighter shading in the deepest
part of the tubercle represents the oldest part which is
caseating; it surrounds an artery which is seen in section.

After very severe choroiditis, or extensive hæmorrhage, the absorption may be incomplete; in addition
to atrophy, we then see grey or white patches, or

lines which in texture and pattern remind us of scars in the skin, or of patches and lines of old thickening on serous membranes.

Very characteristic changes are seen after rupture of the choroid from sudden stretching caused by blows on the front of the eye. These ruptures are never situated far from the disc and yellow spot, and occur in the form of long, tapering lines, fissures of atrophy, usually curved slightly towards the disc, and sometimes branched; their borders are often pigmented. If seen soon after the blow the rent is more or less hidden by blood, and the retina over it is hazy.

The pathological condition known as "colloid disease" of the choroid consists in the growth of very small protuberances from the thin *lamina elastica*, which underlies the pigment epithelium. It is common in eyes excised for old inflammatory mischief, and in partial atrophy after choroiditis (Fig. 39). But little is known of its ophthalmoscopic equivalent, or its

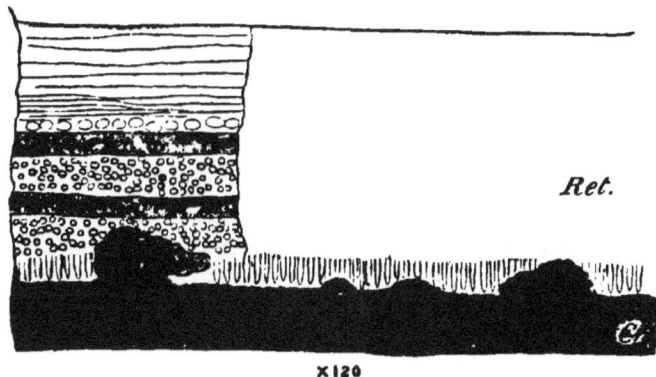

× 120

FIG. 39.—Partial atrophy after syphilitic choroiditis. Minute growths from inner surface of choroid, showing how they disturb the outer layers of the retina.

clinical characters. Probably it may result from various forms of choroiditis, and may also be a natural senile change.

Hæmorrhages from the choroidal are not so often recognised as those from the retinal vessels, but may be seen sometimes, especially in old people and in highly myopic eyes. They are more abruptly defined and more rounded than retinal hæmorrhages, and it is generally possible to recognise the striation of the overlying retina. Occasionally they are of immense size.

In old persons a form of choroidal disease is met with in which, along with superficial atrophy, the large deep vessels are much narrowed or even converted into white lines quite devoid of blood column, by thickening of their coats (atheroma). This change is seen oftenest near the disc.

Clinical forms of choroidal disease

(1.) Numerous discrete patches of choroidal atrophy (sometimes complete, as if a round bit had been punched out; in others incomplete, though equally round and well defined) are scattered about in different parts of the fundus, but are most abundant towards the periphery; or, if scanty, are found only there. They are more or less pigmented, unless the patient's complexion is extremely fair (Fig. 36).

(2.) The disease has the same distribution, but the patches are confluent; or large areas of incomplete atrophy, passing by not very well-defined boundaries into the healthy choroid around, are interspersed with a certain number of separate patches; or there may be no separate patches, but a widely-spread superficial atrophy with pigmentation may occupy a large part of the fundus (Fig. 35, upper part).

These two types of *choroiditis disseminata* run into one another, several different names being used by authors to indicate topographical varieties. Generally both eyes are affected, though not quite equally, and in some cases only one. The retina and disc often, but not always, participate in the inflammation.

Syphilis is almost invariably the cause of symmetrical disseminated choroiditis. The choroiditis generally occurs from one to three years after the primary disease, whether this be acquired or inherited, and seldom later than five years. In a few cases it sets in later.

The discrete variety (Fig. 35, lower part), where the patches, though usually involving the whole thickness of the choroid, are not connected together by areas of superficial change, is the less serious form, unless very large patches are affected. A moderate number of such patches confined to the periphery cause no appreciable damage to sight.

The more superficial and widely-spread varieties, in which the retina and disc are implicated from the first, are far more serious. The capillary layer of the choroid seldom again becomes healthy, and with its atrophy, even if the deeper vessels are not much changed, the nutrition of the retina suffers. The retina passes into a condition of slowly progressive atrophy, often with pigmentation in lace-like patterns and spots (Fig. 41), with a peculiar yellowish and slightly hazy atrophy of the disc (" choroiditic atrophy,"—Gowers), and great shrinking of the retinal blood-vessels. The state of the retina often finally becomes very like that in true retinitis pigmentosa, and the patient, as in that disease, suffers from marked night blindness. Such cases continue to get worse for many years, and in the end may become practically blind.

Syphilitic choroiditis generally gives rise, at an early date, to opacities in the vitreous; they are either of large size, and easily seen as slowly floating ill-defined clouds, or so minute and so numerous as to cause a diffuse and somewhat dense haziness, no opacities being separately visible (see p. 195). Some of the larger ones may be permanent. In the advanced stages, as in true retinitis pigmentosa, posterior polar cataract is sometimes developed.

There are no constant differences between choroiditis in acquired and in inherited syphilis; in

11

many cases it would be impossible to guess, from the ophthalmoscopic changes, with which form of the disease we had to do.   But there is, on the whole, a greater tendency towards pigmentation in the choroiditis of hereditary than in that of acquired syphilis, and this applies both to the choroidal patches and to the subsequent retinal pigmentation.

In the treatment of syphilitic choroiditis we rely almost entirely on the constitutional remedies for syphilis—mercury and iodide of potassium.   Cases which are treated early in the exudation stage are very much benefited in sight by mercury, and the visible exudations melt away very quickly; but I believe that even in these, complete restitution seldom takes place, the nutrition and arrangement of the pigment epithelium and bacillary layer of the retina being quickly and permanently damaged by exudations into or upon the chorio-capillaris (as in Fig. 37). In the later periods, when the choroid is thinned by atrophy, or its inner surface roughened by little outgrowths into the retina (Fig. 39), or adhesions and cicatricial contractions have occurred between the two membranes, nothing can be done.   A long mercurial course should, however, always be tried if the sight is still failing rather quickly, even if the changes all look old; for in some cases exacerbations of failure take place from time to time, and internal treatment has a marked influence for good. It is well to prescribe, in addition to internal remedies, rest of the eyes in a dark room, and the employment of the artificial leech or of dry cupping at intervals of a few days, for some weeks.   But in many cases it is difficult to ensure such functional rest, for the patients seldom have pain or other discomfort.

(3.) The choroidal disease is limited to the region of the disc and yellow spot, the central region. There are many varieties of these localised changes.

In *myopia* the elongation which occurs at the

posterior pole of the eye very often causes atrophy of the choroid contiguous to the disc, and usually only on the side next the yellow spot (p. 253). The term posterior staphyloma is used for this form of disease when the eye is myopic, because the atrophy is a sign of posterior bulging of the sclerotic. The term " sclerotico-choroiditis posterior " is also used. A similar, but narrow and less conspicuous, crescent or zone of atrophy around the disc is seen in some other states without myopia, notably in old persons, and in glaucoma. Separate round patches of complete atrophy ("punched-out" patches), at the central region, occasionally accompany the common myopic changes; they are due to the myopia, and not to syphilitic choroiditis. More commonly in myopia ill-defined partial atrophy is seen about the yellow spot, sometimes with splits or lines running horizontally towards this part from the disc.

Large patches of disease at the yellow spot itself, with much accumulation of pigment, are seen in a few cases, and are, no doubt, generally signs of former choroidal hæmorrhage. They cause great central defect of vision. They are commonest in old people. Less marked changes, chiefly of the epithelium, but obvious enough, may follow the absorption of large retinal hæmorrhages and of the white patches in albuminuric retinitis.

In another form of disease we find the central region occupied by a number of very small white or yellowish-white dots, sometimes visible only in the erect image. This form, which in typical cases is very peculiar, is seldom seen excepting in old people, and appears to be almost stationary. The discs are often decidedly pale. When very abundant the spots coalesce, and a certain amount of pigmentation is found. The pathological anatomy and general relations of this disease are incompletely known; it has been clinically described by Hutchinson and Tay, and I have seen many examples of it. It is symmetrical, and the changes may sometimes be mistaken for a slight albuminuric retinitis (see p. 178). There is often incipient cataract, and every cataract case should, when possible, be examined for these choroidal changes. No treatment seems to have any influence.

Single spots of choroidal atrophy of small size, and especially towards the periphery, should, no less than abundant changes, always excite grave suspicion of former syphilis, and often furnish valuable corroborative evidence of that disease. The periphery cannot be fully examined unless the pupil be widely dilated.

A few small scattered spots of black pigment on the choroid or in the retina, without evidence of atrophy of the choroid, often indicate former hæmorrhages. They are seen after recovery from albuminuric retinitis with hæmorrhages, and occasionally after blows on the eye; such spots may perhaps sometimes be the result of hæmorrhages from violent retching.

(4.) *Anomalous forms of choroidal disease.*—Single, large patches of complete atrophy, with pigmentation and not located in any particular part, are occasionally met with. There is reason to believe that some of them have followed the absorption of tubercular growths in the choroid, while others are the result of single large hæmorrhages (p. 160). Single large patches of exudation are also met with, and are perhaps tubercular. *Generalised choroidal disease* in patches sometimes occurs in persons who have certainly not had syphilis. I believe that in most of these the disease is due to numerous scattered hæmorrhages into the choroid, sometimes occurring repeatedly at different dates, and leading to patches of partial atrophy with pigmentation. The local cause of the hæmorrhage is obscure and the disease often affects only one eye; the cases are generally seen in young adults, and are not peculiar to either sex.

They may perhaps be called hæmorrhagic choroiditis (compare p. 196 (4)). Although the changes produced are very gross, some of these patients regain almost perfect sight, a fact, perhaps, pointing to the deep layers of the choroid as the seat of disease. External signs of inflammation are sometimes present, the onset of the disease, or its exacerbations, being sometimes accompanied by considerable pain of a neuralgic kind and by some redness of the eyeball ; but these same symptoms may also be met with in syphilitic choroiditis.

Congestion of the choroid is sometimes spoken of as a condition recognisable by the ophthalmoscope, but unless variations of tint or shade are proved to occur at different times in the same patient, the term as an ophthalmoscopic one is worthless.   That active congestion does occur is certain, and it would seem that myopic eyes are especially liable to it, particularly when exposed to bright light and great heat.   Serious hæmorrhage may undoubtedly be excited under such circumstances.   In conditions of extreme anæmia the whole choroid becomes unmistakeably pale.

Coloboma of the choroid (congenital deficiency of the lower part, from the optic nerve to the ciliary body) is shown ophthalmoscopically by a large surface of exposed sclerotic embracing the disc (which is much altered in form, and may be hardly recognisable), and extending downwards quite to the periphery, where it often narrows to a mere line or chink. The surface of the sclerotic, as judged by the course of the retinal vessels, is often very irregular, from bulging of its floor backwards.   The coloboma is occasionally quite small, and limited to the part around the nerve.   Coloboma of the choroid is seldom, if ever, seen without coloboma of the iris, though the two are not always of proportionate size.

Albinism is accompanied by congenital absence of pigment in the cells of the pigment epithelium and stroma of the whole uveal tract (choroid, ciliary processes, and iris).   The pupil looks pink because the fundus is lighted to a great extent indirectly through the sclerotic.   Sight is always defective, and the eye photophobic and usually oscillating.   Many almost albinotic children become moderately pigmented as they grow up.

CHAPTER XIV

DISEASES OF THE RETINA

Of the many morbid changes to which the retina is subject, some begin and end in this membrane, such as albuminuric retinitis, and many forms of retinal hæmorrhage; in others, the retina takes part in changes which begin in the optic nerve (neuro-retinitis), or in the choroid (choroido-retinitis); very serious lesions also occur from embolism or thrombosis of the central retinal vessels. The retina may be separated (" detached") by blood or other fluids from its attachment to the choroid; and it may also be the seat of malignant growth (glioma), and probably of tubercular inflammation.

In health the human retina is so nearly transparent as to be almost invisible by the ophthalmoscope during life, or to the naked eye if examined immediately after excision. We see the retinal blood-vessels, but the retina itself, as a rule, we do not see. The main blood-vessels are smaller and much less abundant than those of the choroid; the veins and arteries in general terms form pairs, the veins not being more numerous than the arteries. All the vessels come from or to the optic disc. At the disc capillary anastomoses are formed between the vessels of the retina and those of the choroid and sclerotic. But as no other anastomoses are formed by the vessels of the retina, the retinal circulation beyond the disc is terminal; and, further, as the vessels branch dichotomously and the branches anastomose only by means of their capillaries, the circulation of each considerable branch is terminal also. The retinal

vessels are derived from the arteria and vena centralis,
which run in the trunk of the optic nerve; it is,
however, quite common for one or more small
branches to be derived from choroidal or scleral vessels
which run into the disc and then turn forwards and
pass into the retina; but these, like the rest, are
terminal.   The capillaries, which are not visible by
the ophthalmoscope, are narrower and much less
abundant (excepting just at the yellow-spot region)
than those of the choroid (compare Figs. 34 and 40),
their meshes becoming wider and wider towards the
anterior and less important parts of the retina.
They are most abundant at the yellow-spot region, the
only part used for accurate vision; the very centre of
this region (*fovea centralis*), however, where all the
layers excepting the cones and outer granules are excessively thin, contains no vessels, the capillaries
forming fine close loops just around it (Fig. 40).

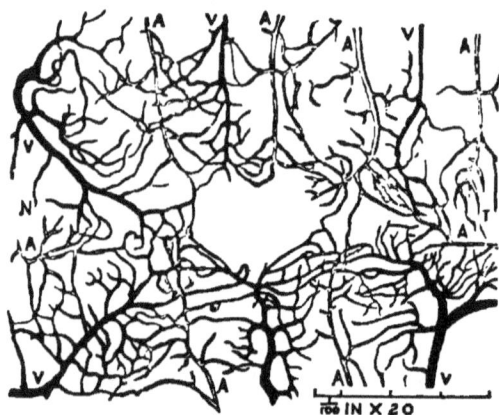

FIG. 40.—Blood-vessels of human retina at the yellow spot. The
central gap corresponds to the fovea centralis. A. Arteries;
v. Veins; N. Nasal side (towards disc); T. Temporal side.

In children, especially those of dark complexion, a
peculiar and striking whitish shifting reflexion, or
shimmer, is often seen at the yellow-spot region and
along the course of the principal vessels.   It changes

with every movement of the mirror, and reminds one of the shifting reflexion from " watered " and " shot " silk. Around the yellow spot it takes the form of a ring or zone, and is known as the " halo round the macula " (p. 34).

When the choroid is highly pigmented, even if this shifting reflexion is absent, the retina is visible as a faint haze over the choroid like the " bloom " on a plum. Under the high magnifying power of the erect image the nerve-fibre layer of the retina is often visible near the disc, as a faintly marked radiating striation ; whilst the sheaths of the large central vessels at their emergence from the physiological pit (p. 33) show many variations in thickness and opacity.

In rare cases the medullary sheath of the optic nerve-fibres, which should cease at the lamina cribrosa, is continued up to, or reproduced at the disc, especially at its margin, and causes the ophthalmoscopic appearance known as " opaque nerve-fibres." This congenital peculiarity may affect the whole circumference of the disc or only a patch or tuft of the fibres ; it may only just overleap the edge of the disc, or may extend far into the retina, where even separate islands of it are sometimes seen. It is to be particularly noted that the central part (physiological pit) of the disc is never affected, because it contains no nerve-fibres. The parts affected are pure white, and quite opaque ; at the margin of the patch, where the change ceases rather gradually, a fine striation is visible, radiating from the disc like finely-carded cotton wool; the retinal vessels are sometimes buried in the opacity, sometimes run unobscured on its surface. The deep layers of the affected part of the retina being obscured by the opacity, an enlargement of the normal " blind-spot " is the result. It may affect one or both eyes. There is seldom any difficulty in distinguishing this condition from opacity due to neuro-retinitis.

Ophthalmoscopic signs of retinal disease

Congestion.—No amount of capillary congestion whether passive or active alters the appearance of the retina; and as to the large vessels, it is better to speak of the arteries as unusually large or tortuous, or of the veins as turgid, or tortuous, than to use so vague a term as congestion. Capillary congestion of the optic disc may undoubtedly be recognised; but even here great caution is needed, and much allowance must be made for differences of contrast depending on the depth of tint of the choroid, for the patient's health and age, and for the brightness of the light used, or, what is the same thing, for the size of the pupil. Caution is also needed against drawing hasty inferences from such slight haziness of the outline of the disc, as may often be seen in cases of hypermetropia, and which is due to accentuation of that natural streakiness caused by the nerve-fibres which has been already referred to.

The only ophthalmoscopic evidence of *active retinitis* is loss of transparency of the retina, and two chief types are soon recognised according as the opacity is diffused, or consists chiefly of abrupt spots and patches. Hæmorrhages are present in many cases of retinitis; but they are also common in cases where there is no true inflammation. The state of the disc is subject to much variation, but it seldom escapes entirely in extensive or prolonged retinitis. In a large majority of cases of recent retinitis the visible changes are limited to the central region where the retina is thickest and most vascular.

(1.) The lessened transparency which accompanies diffused retinitis simply dulls the red choroidal reflex, and the term " smoky " is fairly descriptive of it. The same effect is given by slight haziness of any of the anterior media, but a mistake is excusable only when there is diffused mistiness of the

vitreous from opacities which are too small to be
seen separately (p. 195), and the difficulty is then
increased because this very condition of the vitreous
often coexists with retinitis. A comparison of the
erect and inverted images is often useful. If the
vitreous is foggy the details of the fundus will be
obscured as much or more by the erect image as by
indirect examination, because the illumination is often
less in the erect image; but if the diffused haze
noticed by indirect examination is caused by retinitis,
then the erect image will often give a different impression, because the parts are now much more highly
magnified, and what before seemed a uniform haze
may now show well-marked spotting, or stippling,
or streaking. When the change is pronounced enough
to cause a decidedly white haze of the retina there is
no longer any doubt. The retinal arteries and veins
may be decidedly enlarged and tortuous in retinitis,
and in severe cases they are generally obscured in
some part of their course. These diffused forms are
usually caused either by syphilis or embolism.

(2.) The retina generally is clear, but near the
yellow spot a number of small, intensely white
rounded spots are seen, either quite discrete or partly
confluent. When very abundant and confluent they
form large, pure white, abruptly outlined patches,
often with crenated borders; or some parts may be
striated and others stippled.

(3.) A number of separate white patches are
scattered about the central region, but without
special reference to the yellow spot. They are of
irregular shape, quite white, but perhaps striated
in parts; they are easily distinguished from patches
of choroidal atrophy (p. 157) by their colour, the
comparative softness of their outlines, and the absence
of pigmentation.

In the last two forms, hæmorrhages are usually
present also. Both are generally associated with
albuminuria, but in rare cases similar changes are

caused by cerebral disease. The changes are almost always symmetrical.

(4.) There are numerous hæmorrhages with general haziness intensified at places into distinct patches of white or yellowish white, but not abruptly defined ; the retinal vessels are extremely tortuous, and the veins dilated.

(5.) Rarely a single large patch or area of white opacity is seen with softened, ill-defined edges, any retinal vessels that may cross it being obscured. In most cases such a patch of retinitis is caused by choroidal exudation beneath (p. 158).

Hæmorrhage may take place into any of the retinal layers, and the shape of the blood patches is in great part determined by their position. When blood is effused into the nerve-fibre layer, or is confined by the sheath of a large vessel, it takes a linear or streaked form and structure, following the direction of the nerve fibres ; extravasations in the deeper layers are generally rounded or irregular. Very large hæmorrhages, many times as large as the disc, sometimes occur near the yellow spot, and probably all the layers then become infiltrated, while sometimes the blood ruptures the anterior limiting membrane and passes into the vitreous.

Retinal hæmorrhages may be large or small, single or multiple ; limited to the central region or scattered in all parts ; linear, streaky or flame-shaped, punctate or blotchy; they may lie alongside of large vessels, or be in no apparent relation to visible vessels. The hæmorrhage may, as already mentioned, be the primary change or may only form part of a retinitis or neuro-retinitis. A hæmorrhage which is mottled and of dark dull colour is generally old. The rate of absorption varies very greatly ; hæmorrhages after blows are very quickly absorbed, while those depending on rupture of diseased vessels in old people or accompanying albuminuric retinitis, generally last for months, and leave permanent traces (*see* below).

Pigmentation of the retina has been referred to in connection with choroiditis (p. 156). Whenever pigment in the fundus forms long, sharply-defined lines, or is arranged in a mossy or lace-like or reticulated pattern, we may always safely infer that it is situated in the retina, and generally that it is distributed along the sheaths of the retinal vessels (compare Figs. 41 and 40). Pigment in or on the choroid never takes such a pattern; but pigment so

FIG. 41.—Study of pigment in the retina in a specimen of secondary retinitis pigmentosa, seen from the front (vitreous) surface.

arranged is often mingled with round spots or blotches, particularly in cases of choroiditis with secondary affection of the retina. It is true that rounded pigment spots are often situated in the retina, even when no linear or branched figures are present; but such rounded spots in the retina are indistinguishable during life from similar spots on the choroid. In every case where we decide that the retina is pigmented the choroid must be carefully examined for evidences of former choroiditis.

Spots of pigment are not unfrequently left after the absorption of retinal hæmorrhages. It is not often difficult to distinguish these spots from the results of choroiditis; they are uniformly black or dark brown, and though sometimes surrounded by a little collar of pale choroid or by some disturbance of the pigment epithelium they are not associated with any other indications of choroidal disease (compare Choroidal Hæmorrhages, pp. 160 and 164).

Atrophy of the retina, of which pigmentation, when present, is always a sign, has for its most constant indication a marked shrinking of the retinal blood-vessels and thickening of their coats. When the atrophy follows a retinitis or choroido-retinitis (retinitis pigmentosa, syphilitic choroido-retinitis, &c.) all the layers are involved, and the outer layers (those nearest the choroid) before the inner; but when it is secondary to disease of the optic nerve (optic neuritis, progressive atrophy, and glaucoma) only the layers of nerve-fibres and ganglion cells are atrophied, the outer layers remaining perfect even after many years. A retina atrophied after retinitis often does not regain its full transparency, and if there have been choroiditis the retina remains especially hazy in the parts where this has been most severe.

The disc in atrophy following retinitis or choroido-retinitis always passes into atrophy, but its appearance is often peculiar; pale, hazy, but homogeneous looking, with a yellowish or brownish tint, sometimes described as " waxy " (Hutchinson. See also p. 161).

Detachment (separation) of the retina.—As there is no continuity of structure between the choroid and

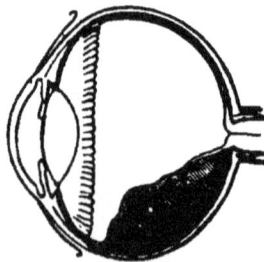

FIG. 42.—Section of eye with partial detachment of Retina.

retina, the two may be easily separated by hæmorrhage, effusion of fluid, and morbid growths. This result is very seldom caused by primary changes in the retina, but nearly always depends upon disease of the choroid or vitreous. The retina is raised up from the choroid (at the expense of the vitreous, which is proportionately absorbed) like a blister, but always remains attached at the disc and ora ser-rata, unless as the result of wound or great vio-lence. The depth and area of the separation are subject to much variety. Fig. 42 shows a diagram-matic section of an eye in which the lower part of the retina is separated.

The separated portion is usually far within the focal length of the eye, and is, therefore, visible with great ease in the erect image (p. 36), when it appears as a dark, or grey, or whitish reflexion in some part of the field, the remainder being of the proper red colour; the detached part is grey or whitish, because either the retina or the underlying fluid is opaque. With care we are able accurately to focus the surface of the grey reflexion, to see that it is folded and to see one or more retinal vessels meandering upon it in a tortuous course, and appearing of small size and dark colour. If the separation is deep the outline of its more prominent folds (Fig. 43) can be seen standing out sharply against the red background, and in

FIG. 43.—Ophthalmoscopic appearance of detached retina (erect image) (after Wecker and Jaeger).

many cases the folds will flap about when the eye is quickly moved. In extreme cases we can see the detachment by focal light. When the detachment is shallow, the red reflex of the choroid is still seen, and the diagnosis then rests on the observation of whether the vessels become darker, smaller, and more tortuous at any part of the fundus, and upon ophthalmoscopic estimate of the refraction of the retinal vessels (p. 37) at different parts of the fundus, for the detached part will be much more hypermetropic than the rest. It should be added that in very high myopia, a shallow detachment may still lie behind the principal focus, and therefore not yield an erect image without a suitable concave lens. In such cases, and in others where minute rucks or folds of detachment are present, examination by the indirect method will always lead to a right diagnosis, the image of the detached portion not being in focus at the same moment as its surrounding parts, and the vessels always being tortuous.

Deep and extensive detachment is often associated with opacities in the vitreous or lens, or with iritic adhesions, all or any of which conditions militate very much against the conclusive application of the above tests, for the full use of which a dilated pupil is often essential. (For the causes of detachment, *see* Injuries, Myopia, Tumours.)

Clinical forms of retinal disease

The symptoms of the different forms of retinal disease relate only to the failure of sight which they cause, and this may be either general or confined to a part of the field, according to the nature of the case. Neither photophobia nor pain occurs in uncomplicated retinitis.

Syphilitic retinitis is generally associated with and secondary to choroiditis (p. 161), but in some cases no change can be demonstrated in the choroid, and the retinitis, though apparently quite the same, is assumed to be primary. The vitreous in this disease, as in syphilitic choroiditis, is often hazy, and the opacities are sometimes seated very deeply, just in front of the retina. The changes are those of diffuse retinitis ((1) p. 170), with slight " smoky " haze, often confined to the yellow spot or disc region, but in bad cases passing into a whiter mistiness, and extending over a much larger region ; sometimes long branching streaks or bands of dense opacity are met with, and hæmorrhages may occur. The disc is always hazy, and at first decidedly too red, while the retinal vessels, both arteries and veins, are somewhat turgid and tortuous. In a few the disc becomes opaque and swollen (neuritis). At a late period in unfavorable cases the vessels shrink slowly, almost to threads, and the retina often becomes pigmented at the periphery.

Syphilitic retinitis is one of the secondary symptoms, seldom setting in earlier than six, or later than eighteen, months after the primary disease. It occurs in congenital as well as acquired syphilis. It generally attacks both eyes, though often with a short interval. Its onset is often rapid, as judged by its chief symptom, failure of sight, and it may be stated that, as a rule, the degree of amblyopia is much greater than would be expected from the comparatively slight visible changes. It is essentially a protracted disease, always lasting for months, and showing a remarkable tendency for many months to repeated and rapid exacerbations after temporary recoveries, but with a tendency to get worse rather than towards spontaneous cure. Its onset is sometimes attended by pain. One of the early symptoms is often a " flickering " before the sight, and this with the history of variations lasting for a few days, and in later stages the presence of marked night-blindness, often lead to a correct surmise before ophthalmoscopic examination. There is, however, nothing pathognomonic in any of the symptoms.

Mercury produces most marked benefit, and when used early it permanently cures a large proportion of the cases; but in a number of cases, perhaps in those where there is most choroiditis, the disease goes slowly from bad to worse for several years, in spite of very prolonged mercurial treatment. Of the efficacy of prolonged disuse of the eyes, and of local counter-irritation or depletion, strongly recommended by many authors, I have had but little experience. In the rebellious cases it might be worth while to try the subcutaneous administration of mercury.

Albuminuric retinitis (neuro-retinitis).—The changes are strongly marked, and so characteristic that it is possible, in most cases, to say from an ophthalmoscopic examination alone that the patient is suffering from chronic kidney disease.

The earliest change (the stage of œdema and exudation) is a general haze of a dull or greyish tint in the central region of the retina, generally with some retinal hæmorrhages, and with or without haze and swelling of the disc. In this stage there is nothing quite characteristic; the sight is often unimpaired, and so the cases are seldom seen by ophthalmic surgeons till a few weeks later, when the translucent, probably albuminous exudations into the swollen retina have passed into fatty or fibrinous degeneration, affecting both the nerve-fibres and connective tissue of the retina.

In this condition, the second stage, we find a number of pure white dots, spots, or patches, in the hazy region, and especially grouped around the yellow spot; their peculiarity is their pure opaque white colour, which is almost glistening when the spots are small and round. When not very numerous they are almost confined to the yellow spot region, from which they show a tendency to radiate in lines; when very small and scanty they may be overlooked,

unless we employ the erect image; but in most cases
large easily seen patches are formed by the confluence
of small spots, and the borders of these patches are
striated, crenated or spotted. Again, in a few we
find none of the small, round, bright dots, nor any
large confluent areas, but a number of separate white
patches, of moderate size (perhaps a quarter as large
as the disc) and irregular shape ((3), p. 171). In
this stage the disc is generally hazy and somewhat
swollen, especially just at its margin, and the retina,
as judged by the undulations of its vessels, and confirmed by post-mortem examinations, is much thickened. Hæmorrhages are generally present in greater
or less number, and occasionally constitute the most
marked feature of the case; they are usually striated.
Sometimes an artery is seen sheathed by a dense
white coating.*   In another group neuritis (p. 219) is
the most marked change, though some bright white
retinal spots are always to be found by careful
examination.

The natural tendency is towards absorption of the
fatty deposits and extravasations, and return to a
more or less healthy state with improvement of
sight—the third stage, or stage of absoption and
atrophy. In the course of several months the white
spots diminish in size and number until only a few
very small ones are left near the yellow spot, with,
perhaps, some residual haze; the blood-patches are
slowly absorbed, often leaving pigment spots, and
the retinal arteries may be shrunken. In cases of
moderate severity almost perfect sight is restored.
But when the optic nerve suffers severely (severe
neuritis), or if the retinal disease is excessive and
attended by great œdema, sight either improves very
little, or, as the disc and retina pass into atrophy
and the retinal vessels shrink, almost total blindness
comes on. Such a condition may easily be mistaken

* An excellent illustration of this is given in Dr. Gowers'
'Medical Ophthalmoscopy,' pl. xii, fig. 1.

for atrophy after cerebral neuritis ; but the presence of a few minute bright dots at the yellow spot, or of some scattered pigment spots left by extravasations, or of some superficial disturbance of the choroid near the yellow spot, will generally lead to a correct inference (p. 173).

In the cases attended by the greatest swelling and opacity of retina and disc ("inflammatory form") death often occurs before retrogressive changes have taken place (Gowers).

Albuminuric neuro-retinitis is always symmetrical, but seldom quite equal in degree or in final result in the two eyes. In extreme cases it causes detachment of the retina. In rare cases the renal disease causes choroiditis and iritis.

The kidney disease in the malady under consideration is always chronic. The retinitis may occur in any chronic nephritis, and in the albuminuria of pregnancy. Whatever be the form of the kidney disease, the retinitis seldom occurs without other signs of active kidney mischief, such as headache, vomiting, and loss of appetite, and often anasarca. The quantity of albumen varies very much. In the absence of anasarca the symptoms are often put down to "biliousness," and such persons seek advice first on account of their failing sight, when the ophthalmoscope leads to the correct diagnosis. Many of the best marked cases of albuminuric retinitis occur in the albuminuria of pregnancy, and the prognosis for sight is good in many of these if the symptoms come on late in the pregnancy. On the other hand, some of them (probably cases of old kidney disease) do very badly, and pass into atrophy of the nerves. A second attack of retinitis sometimes occurs in connection with a relapse of renal symptoms.

(For the changes which occur in the retina in some other chronic general diseases, diabetes, pernicious anæmia, and leucocythæmia, see Chapter on Etiology).

The term *Retinitis hæmorrhagica* has been given to

certain rare cases, where very numerous small linear or flame-shaped hæmorrhages, situated, therefore, in the nerve-fibre layer, are found all over the fundus, usually in only one eye, and with rapid onset. The patients are often gouty. Thrombosis of the trunk of the vena centralis retinæ has been suggested as the determining cause of the bleeding.*

Other cases are seen where extravasations, varying much in size, number, and shape, are scattered in different parts of the fundus of one or both eyes. Some of them are probably allied to the above, but often the nature of the case is obscure, or the hæmorrhages are related to senile degeneration of vessels. Such cases are often called *retinitis apoplectica*.

Lastly, in an important group a single very large extravasation occurs from rupture of a large retinal vessel, probably an artery. The hæmorrhage generally occupies the yellow spot region; in process of absorption it becomes mottled, the densest parts remaining longest, and, if seen in that condition for the first time, the case may be taken for one of multiple hæmorrhages. These large hæmorrhages cause great defect of sight, which comes on in an hour or two, but not with absolute suddenness. Absorption, in all the groups of cases just mentioned, is very slow.

Hæmorrhages may occur from blows on the eye. They are usually small, and quickly absorbed, differing in the latter respect very much from the cases before described.

Embolism of the central artery of the retina, or of one or more of its main divisions, gives rise to a retinitis so peculiar that its cause can in most cases be recognised at once if it be recent; whilst in old cases the appearances, taken with the history, always lead to a right diagnosis.

The leading symptom is the occurrence of an instantaneous defect of sight, which the patient finds

* Hutchinson.

on trial to be seated in one eye, or sometimes he thinks that one eye has suddenly become "shut," the blindness being as sudden as that from quickly closing the lids; but whether the defect amounts to absolute blindness or not, depends on the position and size of the plug, and on other circumstances. In any case, owing to the temporary establishment of collateral circulation by the capillary anastomoses at the disc (p. 167), the patient often notices an improvement of sight a few hours after the occurrence. But this improvement is only very slight, for the collateral channels are utterly insufficient to meet the demand quickly; nor is it often permanent, because the retina suffers very quickly from the almost complete stasis, œdema and inflammation rapidly setting in, and leading to permanent damage. The changes are always greatest in the thickest, most vascular, and functionally most important part, viz. around the yellow spot and near the disc.

If the case is seen within a few days of the occurrence the red reflex of the choroid around the yellow spot and disc is quite obscured, or partially dulled, by a diffused and uniform white mist. The opacity is greatest just around the centre of the yellow spot, where the retina is very vascular (Fig. 40), and where its cellular elements (ganglion and granule layers) are more abundant than anywhere else; but at the very centre of the white mist a small, round, red spot is seen, so well defined that it may be mistaken for a hæmorrhage; it is really the *fovea centralis*, where the retinal layers are so thin that the choroid continues to shine through it when the surrounding parts are opaque; it is spoken of by authors as the "cherry-red spot at the macula lutea." This appearance is never seen except after sudden arrest of arterial blood supply, as by embolism or thrombosis of the arteria centralis, or perhaps sudden hæmorrhage into the optic nerve compressing the vessels, and of these causes embolism appears to be the commonest. The haze surrounds and generally affects the disc also, the latter being too red if seen within a day or so of the occurrence, but soon becoming pale. The small veins in the yellow-spot region often stand out with great distinctness, partly because enlarged by stasis, and partly from contrast with the white retina on which they lie. Small hæmorrhages are often present in the hazy retina. The larger retinal vessels, both arteries and veins, are more or less diminished at and near the disc, the arteries in the most typical cases being reduced to mere threads, while both arteries and veins are sometimes observed to increase in size as they recede from the disc. The retinal arteries are not always extremely shrunken in cases of embolism, the variations depending upon the position and size of the plug, upon whether it causes complete occlusion or not. The sudden and complete failure of supply to a retinal arterial branch is sometimes followed by its emptying and shrinking to a white cord almost immediately, whilst in other cases a large artery may be little, if at all, altered in size, although its blood column is quite stagnant, as is proved by the impossibility of producing pulsation in it by the firmest pressure on the globe, whilst the other branches respond perfectly to this test. In other cases, again, this pressure test, which showed blockage of some or all branches shortly after the onset, indicates perfectly restored communication a few days later, without any visible evidence of collateral circulation.

In from one to about four weeks the cloudiness clears off, and the disc passes into moderately white atrophy. The arteries, or some of them (according to the position of the plugging), will perhaps be easily seen as white lines instead of blood columns ; or they may be only considerably reduced in size, but still pulsating easily on pressure.

Sight is always extinguished, or only perception of large objects remains, whatever be the final state of the blood-vessels. In the rare cases, where a single embolus passes beyond the disc, and is arrested in a branch at some distance from it, the changes are confined to the corresponding section of the retina, and a limited defect of the field is the only result.

It is scarcely necessary to say that no treatment can be of any use in cases of lasting occlusion of the retinal arteries.

In certain rare cases where instantaneous blindness of both eyes has been associated with extremely diminished arteries ("ischæmia retinæ"), iridectomy has been followed by return of sight; lower tension causing re-establishment of circulation. These cases generally occur after whooping-cough.

Retinitis pigmentosa is a very slowly progressive symmetrical disease, leading to atrophy of the retina with great contraction of its vessels, and accompanied by the collection of black pigment in its layers and around the blood-vessels, and by secondary atrophy of the disc. Slowly progressive deafness sometimes accompanies the disease.

The earliest symptom is inability to see well at night, or by artificial light (night-blindness, hemeralopia) and contraction of the visual field ; and these defects may reach a high degree, whilst central vision remains excellent in bright daylight. The symptoms are noticed at an earlier stage by patients in whom the choroid is dark and absorbs much light.

Ophthalmoscopic examination in a case where these symptoms have been present for some years shows the following changes :—(1) At the equator or periphery a greater or less quantity of pigment arranged in a reticulated or linear manner (Fig 41), often mingled with small dots ; (2) in advanced cases there is evidence of removal of the pigment epithelium, but never any patches of choroidal atrophy; (3) the pigment is always arranged in a belt, which is in general terms uniform, the pigment being most crowded at its centre and thinning out towards its anterior and posterior borders; (4) the changes are always symmetrical, and the symmetry is very precise. These appearances are quite characteristic of true retinitis pigmentosa. In addition we find (5) diminution in size of the retinal blood-vessels, the arteries in advanced cases being mere threads; (6) a peculiar hazy, yellowish, "waxy" pallor of the optic disc (p. 174); (7) sometimes the pigmented parts of the retina are very decidedly hazy. These latter changes, however, are found in many cases of late retinitis consecutive to choroiditis, and are not peculiar to the present malady; (8) posterior polar cataract and disease of the vitreous are often present in the later stages.

The disease generally progresses slowly to almost complete blindness; but it may come to a standstill after progressing to a certain point. The quantity of pigment visible by the ophthalmoscope is very different in different cases of about equal duration, and is not in direct relation to the defect of sight. Cases occur which certainly belong to the same category in which no pigment is visible during life, the retina being merely hazy, and in one such case microscopical examination revealed abundance of minutely divided pigment (Poncet).

The pathogenesis of the disease is not finally settled; it is at present doubtful whether there is from the first a slow sclerosis of the connective-tissue elements of the retina, with passage inwards of pigment from the pigment epithelium, or whether the disease begins in the superficial layer of the choroid and the pigment epithelium.

It generally begins in childhood or adolescence; but a few cases of apparently recent origin are seen in quite aged persons, and a few are considered to be truly congenital. Its cause is obscure. It is undoubtedly strongly heritable, and many excellent observers believe that it is really produced by consanguinity of marriage, either between the parents, or between near ancestors, of the affected person. Many of its subjects are of full mental and bodily vigour; but others are badly grown and intellectually defective. Although want of education, as a consequence of defective sight and hearing, may sometimes account for this result, we cannot thus explain the various defects and diseases of the nervous system which are not unfrequently noticed in kinsmen. That the subjects of this disease should be discouraged from marrying is sufficiently evident, for there is no doubt that it is strongly heritable.

No treatment has any influence on its course.

Complications such as cataract and myopia are not uncommon and must be treated on general principles, but it is very rarely that we have to extract a cataract.

There are cases in which great difficulty is experienced in distinguishing widely-diffused and superficial choroiditis, with secondary pigmentation of retina and waxy atrophy of the disc, from true retinitis pigmentosa. The question will generally relate to cause, as between retinitis pigmentosa proper and choroido-retinitis from syphilis (especially inherited syphilis) (p. 161). But other cases of choroido-retinal disease occur, which though easily distinguishable from retinitis pigmentosa are, like it, related to some general disease of the nervous system in the patient or his parents, and are not due to syphilis.

FUNCTIONAL DISEASES OF THE RETINA

Functional night blindness (endemic hemeralopia, moon-blindness) is caused by temporary exhaustion (partial anæsthesia) of the retina from prolonged exposure to diffused bright light.

The circumstances under which it comes on are always such as imply not only great exposure to light, but lowered nutrition of the system. Possibly its actual cause may be some constant defect in diet,

such as want of fresh meat or vegetables. Sleeping
with the face exposed to bright moonlight is believed
to help in its production. Thus it is commonest in
sailors after a long voyage with hardships in the
tropics, and in soldiers after long marching in hot
countries (bright sun), and is occasionally seen in
large schools and poorhouses even in our own country. In some countries it is prevalent every year in
Lent, when no meat is eaten, and again in har-
vest time. The disease is rare in our country,
the sporadic cases that come under notice being
generally in sailors just returned from sea, often
with a history of scurvy ; but scattered cases occur
in the resident population, usually in children.
A peculiar change is sometimes seen in the presence of two little dry patches on the conjunctiva, at the inner and outer border of the cornea.
They consist of a film of fatty or sebaceous matter,
with epithelial scales, and are said sometimes to
contain air-bubbles. Whatever may be the mode of
their production, they are not essential to this disease, being seen occasionally in other conditions.
This form of night blindness causes no ophthalmoscopic changes. It is always quickly cured by protection from bright light and by improvement of the
general health. The efficacy of protecting the retina
from the exhausting effect of light is shown by the
fact that bandaging one eye during the daytime has
been found to restore its function sufficiently for useful
sight during the ensuing night's watch, the unprotected fellow-eye remaining as defective as ever.

Snow-blindness (or *ice-blindness*) appears to be
essentially the same disease, with the superaddition of
conjunctival congestion, throbbing pain, and photophobia, and sometimes of ecchymoses into the conjunctiva. These changes are thought to depend, in
some measure at any rate, on the effect of the rarefied
atmosphere in which, in mountaineering cases, the
exposure occurs. It may be suggested that the

local effect of the intense cold on the conjunctiva may take a large share in the result by paralysing the vessels, and leading to their over-distension as soon as a warmer climate or warm room is reached. Snow-blindness is effectually prevented by wearing dark-colored glasses.

Under the name of *Hyperæsthesia of the retina*, or of the optic nerve, cases are described in which, together with a varying degree of intolerance of light, there is irritability and want of endurance of the ciliary muscle (accommodative asthenopia), with some conjunctival irritability; lastly, in some cases there is a degree of weakness of the internal recti (muscular asthenopia). It must be added that this assemblage of symptoms is often independent of any considerable error of refraction.

Of the three principal factors, undue sensitiveness of the retina, irritability of conjunctiva and cornea, (hyperæsthesia of fifth nerve), and deficient endurance of the ciliary muscle, any one may be more pronounced than the others, although all three are distinctly present in every case. The rather vague term asthenopia, meaning weakness of the eyes under various circumstances, is, with few exceptions, less misleading than any more precise term. The mutual relations of the optic nerve, the fifth nerve, and the motor nerves of the ciliary muscle to each other are complex, and we cannot go into greater detail here than to say that, given a certain state of nervous system, which we may describe as impressionable or irritable or hyperæsthetic, over-stimulation of any one of the three will set up a hyperæsthetic state of the other two.

The symptoms complained of are most commonly a certain degree of photophobia and irritability of the eyes from exposure to wind, dust, heat, and other local irritants, with inability to continue their use for long together at any near work (sewing, reading, &c.).

To these are often added a sense of "dazzling" and the presence of muscæ, also pain at the back of the eyes, frontal headache, and sometimes neuralgic pain in other parts supplied by the first division of the fifth nerve. The symptoms are generally worse on first waking in the morning, and again after the day's work, and are liable to vary much with the state of the patient's health. Such symptoms, often lasting for months, give rise to great discomfort, and often to serious apprehensions on the part of the patient and his friends that he will become blind, or that he must give up his occupation.

Causation.—A morbid sensitiveness of the nervous system is always present. The patients are seldom either children or old people. The majority are women, either young or not much past middle life, often markedly hysterical, and at the same time of feeble circulation. If men they are emotional, apt to attach great importance to minute symptoms of all kinds, and often somewhat hypochondriacal. To such predispositions some local cause is generally added, and the commonest is close application either at needlework, reading or writing ; sometimes working on bright colours, or at glittering or moving objects, is thought to be especially injurious. In other cases we find only a very moderate use of the eyes, but get a history of some ophthalmia (generally phlyctenulæ or superficial ulcers) years ago, which seems to have left the fifth nerve in an unstable condition.

Treatment.—The refraction should always be carefully tested, and any error corrected by suitable spectacles, which may often with advantage be smoke-coloured ; plain coloured glasses are generally useful if no error of refraction exists. Glasses, however, will not cure the disease, and we must be on our guard against promising too much benefit from their use. The next thing is to assure the patient that there is no ground for serious alarm, and that the symptoms, if not immediately curable, will gradually lessen in severity, and finally cease ; that in the mean time he must follow such general directions as will alleviate him most, must abstain from thinking about his eyes, but that he need not be idle.   The next local measures will be to have a good steady artificial light (not flickering), and shade it so as to keep the light and heat from striking directly on the eyes, for it seems that heat is often as efficient as light in bringing on the symptoms in the evening.   The free use of cold water to the eyes, or the more sparing employment of various weak astringent lotions are often useful.   The patient will always tell us that he is much better after resting the eyes for a day or two, and this fact must be utilised by encouraging out-of-door exercise, and only moderate use of the eyes. General measures must be taken according to the indications, especially in reference to any ovarian, uterine, or digestive troubles, or to sexual exhaustion in men.

Colour-blindness may be congenital or acquired. In the latter case it is symptomatic of some disease of the optic nerve.   It also occurs as one of the symptoms in certain functional disorders (hysterical amblyopia).

Congenital colour-blindness is not often identified unless specially looked for.   According to recent and extended researches in various countries a proportion varying from about 3 to 5 per cent. of the males are colour-blind in greater or less degree.   There is reason to believe that it is commoner in the lower classes than amongst those who are better educated. These facts show the importance of carefully testing all men whose employment renders good perception of colour indispensable, such as railway signalmen and sailors.   Colour blindness is usually partial, *i. e.* for only one colour, but is occasionally total.   The commonest form is that in which green is confused with various shades of grey and of red (green-blindness, red-green-blindness); blindness for blue and yellow is very rare. The blindness may be incomplete, perception of red, e. g. being merely enfeebled, and bright reds being recognised; or it may be complete for all shades and tints of that colour. Congenital colour-blindness is very often hereditary, but nothing further is known of its cause. It is scarcely ever seen in women. The acuteness of vision (i. e. perception of form) is normal; it always affects both eyes.

The detection of colour-blindness, either congenital or acquired, is easy, if, in making the examination, we bear in mind the two following points already referred to at p. 22. Many persons with perfect colour *perception* have a very imperfect knowledge of the *names* of the various colours, and appear colour-blind if asked to name them. Secondly, the really colour-blind are often ignorant of the fact, having learnt to compensate for their defect by attending to other properties commonly associated with the colour of the objects which most concern them, more particularly differences of *shade*. Thus a signal-man may be colour-blind for red and green, and yet may, as a rule, correctly distinguish the green from the red light, because one appears to him "brighter" than the other. The quickest and best way of avoiding these fallacies has been mentioned at p. 22. Certain standard coloured wools are given to the patient without being named, and he is asked to choose from the whole mass of skeins of wool all that appear to him of nearly the same colour and shade (no two being in reality exactly alike). If, for example, he cannot distinguish green from red he will place the green test skein side by side with various shades of grey and red. Wilful concealment of colour-blindness is impossible under this test if a sufficient number of shades be used.

It may here be noted that the normal visual field is not of the same size for all colours, violet, green, and red having the smallest fields, and blue the largest; next, that with diminished intensity of light, some colours are less easily perceived than others, red being the first to become indistinguishable, and therefore requiring for perception the strongest light, blue persisting longest *i. e.*, being perceived under the lowest illumination. In congenital colour-blindness as we have seen, red-green blindness is the commonest form; and in cases of amblyopia from commencing atrophy of the optic nerve green and red are almost always the first colours to fail, blue remaining last.

CHAPTER XV

DISEASES OF THE VITREOUS

THE vitreous humour is nourished by the vessels of
the ciliary body, of the retina, and of the optic disc,
and probably is indirectly influenced by the state of
the choroid also. In many cases disease of the
vitreous can be proved during life to be associated
with (and dependent on) disease of one or other of
the structures named; it is seldom if ever primary
except as a senile change.

Thus, in connection with various surrounding morbid processes, the vitreous may be the seat of inflammation acute or chronic, general or local, and of
hæmorrhage. It may also degenerate, especially in
old age; its cells and solid parts, undergoing fatty
degeneration, become visible as opacities, whilst its
general bulk becomes too fluid. The only change
which we can directly prove in the vitreous during
life is loss of transparency from the presence of
opacities moving, or more rarely fixed, in it; but
from the freedom and quickness of their movements,
some idea may be formed of the consistence or degree
of fluidity of the main bulk.

Opacities in the vitreous may take the form of
large dense masses, as from abundant recent bleeding,
or of membranes like muslin, crape, " bee's wings "
of wine, bands, knotted strings, or isolated dots;
and they may be either recent, or the remains of long
antecedent exudations or hæmorrhages. Again, the
vitreous may become more uniformly misty, owing to
the diffusion of numberless small points of opacity,
sometimes too minute to be separately visible.

Opacities in the vitreous are usually detected, with great ease, by direct ophthalmoscopic examination at from 10″ to 18″ from the patient, but are generally situated too far forward (*i.e.* too far within the focus of the lens-system) to be seen clearly at a very short distance (p. 36). By asking the patient to move his eye sharply and fully from side to side and up and down, the opacities will be seen against the red ground as dark figures, which continue to move after the eye has come to rest; they are thus at once distinguished from opacities in the cornea or lens, or from dimly-seen pigment spots at the fundus, which move only whilst the eye moves. The opacities in the vitreous move just as solid particles and films move in a bottle after the bottle has been shaken, and the quickness and freedom of their movement in the one case as in the other depends very much on the limpidness or the viscidity of the fluid. Whenever opacities in the vitreous pass across the field quickly and make wide movements we may be sure that the humour is too fluid; and the contrary may be concluded when they move very lazily. In some cases only one or two opacities may be present, and may only come into view now and then. Such moveable opacities obscure the fundus both by direct and indirect ophthalmoscopic examination, in proportion to their size, density, and position; a few isolated dots scarcely affect the brightness of the ophthalmoscopic image.

Opacities may also lie quite in the cortex of the vitreous, and be so attached to the retina or disc as to have no independent movement. These are generally single, are found lying either over or near to the disc, and may be the result either of inflammation or of hæmorrhage; they are often membranous, more rarely globular, and not perfectly opaque. Such an opacity should be suspected when by indirect ophthalmoscopic examination, a localised haze or blurring is seen over some part of the disc or its neighbourhood. It is to be looked for by the direct method with the eye at rest, when by carefully focussing (accommodating) for the particular part which appeared hazy, the opacity will come sharply into view without a correcting lens, the observer being at a greater or less distance according to its depth. If the eye be hypermetropic a convex correcting lens may be necessary, and if considerably myopic a concave. The kind of refraction must therefore be known in order to make this examination properly (pp. 28 and 36). Densely opaque white membranes may also form over the disc or upon the retina, whose nature and situation is diagnosed in the same way.

Diffused haziness of the vitreous causes, in a corresponding degree, dimness of outline and darkening of all the details of the fundus, which look as if they were seen through a thin smoke. The disc, in particular, appears red, without really being so. Very much the same appearances may be due to diffused haze of cornea or of the lens, the presence of which will, of course, have been excluded by focal illumination. There are cases, however, where the quantity of light which reaches and returns from the fundus is but little diminished, and where, nevertheless, no details whatever can be seen, even indistinctly, by the most careful examination. Probably, in such a case, the light is scattered by innumerable little particles, each of which is transparent, so that though very little light is absorbed, it is all distorted and broken up, as in passing through ground glass or white fog, or a partial mixture of fluids of different densities, such as glycerin and water. This appearance is found chiefly in syphilitic choroido-retinitis, in which diffuse inflammation of the vitreous (infiltration with cells) is known to occur. It is not always easy, nor indeed possible, to distinguish with absolute certainty between diffuse haze of the vitreous and diffuse haze of the retina (p. 170).

Crystals of cholesterine sometimes form in a fluid vitreous, and are seen with bright illumination as minute dancing golden spangles when the eye moves about (*sparkling synchysis*). They proportionately obscure the fundus. Large opacities just behind the lens may be seen by focal light in their natural colours, and hæmorrhage may be detected in this way. In rare cases minute growths consisting chiefly of blood-vessels form on the retina and project into the vitreous; they are rather curiosities than of practical importance.

Parasites (*cysticercus*) occasionally come to rest in the eye, and in development penetrate into the vitreous; they are rarely seen in England, but are commoner on the Continent. Very rarely a foreign body may be visible in the vitreous.

The following are the conditions with which disease of the vitreous is most commonly associated:

(1.) Myopia of high degree and old standing; the opacities move very freely, showing fluidity of vitreous, and are sharply defined. They are often the result of former hæmorrhage.

(2.) After severe blows, causing rupture of the choroid or of some vessels in the ciliary body. When recent the blood can often be seen by focal light, and if very abundant it so darkens the interior of the eye, that nothing whatever can be seen with the mirror.

(3.) After perforating wounds the opacity will be blood if the case be quite recent. Lymph or pus in the vitreous at the inner surface of the wound gives a yellow or greenish-yellow colour, easily seen by focal light or even by daylight (p. 115).

(4.) In rare cases large dense opacities from hæmorrhage, occur spontaneously in eyes not myopic. Such cases are serious because relapses occur, and detachment of retina may come on. The subjects are generally young adults, and the cause obscure (compare Choroiditis, p. 164).

In all of the above cases detachment of the retina

is likely to occur sooner or later, and if present the difficulty of diagnosis as between the two conditions may be considerable (p. 176).

(5.) Syphilitic choroiditis and retinitis. There is often diffuse haze, in addition to large slowly floating opacities. The change here is due to inflammation, and the opacities may entirely disappear under treatment.

(6.) Some cases of cyclitis and cyclo-iritis (p. 113).

(7.) In the early stage of sympathetic ophthalmitis. The opacities are inflammatory.

(8.) In various cases of old disease of choroid, usually in old persons and without proof of syphilis. No doubt many of these indicate former choroidal hæmorrhages.

(9.) The vitreous is believed to become repeatedly and quickly hazy in the active stages of glaucoma. The point is difficult to settle clinically, because the cornea and aqueous are nearly always, and the lens often, hazy at the same time ; and the opportunity of making pathological examination of specimens of uncomplicated, recent glaucoma, with active symptoms, scarcely occurs. The vitreous, however, is known to be more viscid than normal in some glaucomatous specimens, and to be partially liquefied in others ; a change of transparency is, therefore, highly probable.

CHAPTER XVI

GLAUCOMA is a peculiar and very serious disease, of which the pathognomonic objective symptom is increased tightness of the eye capsule (sclerotic and cornea), "increased tension;" all the other phenomena peculiar to the disease, depend upon this change. The disease is much commoner after middle life, when the sclerotic begins to get less distensible, than before that period; and it is commoner in hypermetropic eyes, where the sclerotic is thicker, than in myopic eyes, where it is thinned by elongation of the globe.

Glaucoma may be primary, coming on in an eye apparently healthy, or the subject of some disease, such as simple senile cataract, which has no influence whatever on the glaucoma. It may also be secondary, caused by some still active disease of the eye, or by conditions which have been left from some previous disease, such as iritis. It is always important and seldom difficult to distinguish between primary and secondary glaucoma, but now and then it is not easy to say whether some other concurrent process, such as iritis, is its cause or its consequence.

Glaucoma differs in severity and rate of progress from the most acute to the most chronic and insidious form. But in every form it is always a progressive disease, and unless checked by treatment always goes on to permanent blindness. It generally attacks both eyes, though not simultaneously, the interval varying from a few days to several years.

It is customary to speak of primary glaucoma as either acute, subacute, or chronic; and this division, though arbitrary, is useful in practice. But we must remember that many intermediate forms are found, and that the same eye may, at different parts of its history, pass through each of the three conditions. It may, indeed, be here observed that acute and subacute outbursts are generally preceded by a so-called "premonitory" stage, in which the symptoms are not only chronic and mild, but remitting; the intervals between these little attacks becoming shorter and shorter, till at length they become continuous, and the glaucomatous state is fully established. Rapid increase of presbyopia, shown by the need for a frequent change of spectacles, is said to be a common premonitory sign; it is probably often overlooked.

Chronic glaucoma sets in with a cloudiness of sight, a "fog" liable to variations, and often quite clearing off for days, or even weeks ("premonitory stage"). But in some cases, so far as the patient knows, the failure is steady, with no variations or remissions, from first to last. During the attacks of "fog" artificial lights seem surrounded by coloured rings ("rainbows" or "halos"), which are to be distinguished from those due to mucus on the cornea in chronic conjunctivitis. The attacks of fog are often noticed only after long use of the eyes, as in the evening, the sight being much better in the early part of the day. The defect of sight is to be distinguished from that caused by incipient nuclear cataract, or by disease of the optic nerve, syphilitic retinitis, or attacks of megrim. Even when the sight has become permanently cloudy, complete recovery no longer occurring between the attacks, variations still take place and form a marked feature. Usually there is neither pain nor congestion at any time.

If we see the patient during one of the brief early fits of cloudy sight, or after the fog has settled down permanently, the following changes will be found. A greater or less defect of sight in only one eye or unequal in the two and not remedied by glasses; the pupil a little larger and less active than its fellow. The anterior chamber is sometimes shallower, and there may be slight dulness of the front of the eye from steaminess of the cornea, or more often from haze of the aqueous, and some engorgement of the vessels which perforate the sclerotic at a little distance from the cornea. The tension will be increased (usually about +1) and the field of vision may be contracted, especially on the nasal side.

The optic disc will be found normal, pale, or sometimes congested, in early cases; pale and cupped (pp. 203—4) at a later stage. The cupping usually occupies the whole surface, but sometimes takes the form of a central depression, indistinguishable from a large steep-sided physiological cup (p. 33). There may be spontaneous pulsation of the arteries on the disc or of the veins; or arterial pulsation, if not spontaneous, will be produced by very slight pressure on the eyeball. If the case, although quite painless and free from congestion, is of old standing, there will often be greater increase of tension, the cornea will be more dim, the pupil considerably dilated, the lens often hazy, the field of vision greatly contracted, and acuteness of vision extremely defective. In nearly all cases of glaucoma the temporal part of the field (nasal part of the retina) retains its function last; and in advanced cases the patient will often himself say, or show by the position in which he places things, that he sees only in this direction.

An eye in which the above symptoms have set in may progress to total blindness in the course of months or one to two years without a single "inflammatory" symptom, without either pain or redness— *chronic painless glaucoma (glaucoma simplex)*; and since the lens often becomes partially opaque, assuming a greyish or greenish hue, cases of chronic glaucoma are sometimes mistaken for senile cataract.

But more commonly, in the course of a chronic case, we find periods of pain and congestion, with more rapid failure of sight; or the disease sets in with "inflammatory" symptoms at once. Indeed, the commonest cases are those of *subacute glaucoma*, where, in addition to the symptoms enumerated above, we find dusky reticulated congestion of the small and large episcleral vessels in the ciliary region, with pain referred to the eye, to the side of the head, or down the nose, and rapid failure of sight. The increase of tension, steaminess of cornea, dilatation and sluggishness of pupil, and shallowing of the anterior chamber, are all more marked than is usual in chronic cases; and the haze of the media is too great to allow a good ophthalmoscopic examination.

These symptoms, ending after a few weeks in complete blindness, may keep at about the same height for months afterwards with slight variations, the eye gradually settling down into a permanent state of severe but chronic non-inflammatory glaucomatous tension. In other cases a subacute attack passes off only to return in greater severity a few weeks or days later.

Acute glaucoma differs from the other forms only in suddenness of onset, rapidity of loss of sight, and severity of congestion and pain. The congestion, both arterial and venous, is intense; in extreme cases the lids and conjunctiva are swollen, and there is photophobia, so that the case may be mistaken for an acute ophthalmia. All the signs of glaucoma are intensified; the pupil considerably dilated, often oval upwards, and quite motionless to light; the cornea very steamy, the anterior chamber very shallow, and tension + 2 or 3. Sight will fall in a day or two down to the power of only counting fingers, or to mere perception of light, and if the case have lasted a week or two all p. l. may be abolished. The pain is usually very severe in the eye, temple, and back of head and down the nose; not unfrequently it is so bad as to cause vomiting, and the case is often mistaken, even by medical men, for a "bilious attack" with a "cold in the eye," for "neuralgia in the head," or "rheumatic ophthalmia." Some cases, however, whilst inflammatory and very acute, are mild in degree, and undergo spontaneous remissions; but such cases often pass on into the severely acute type just described.

Glaucoma of any variety, which has led to permanent blindness, is called "*absolute glaucoma.*" Such an eye continues to display the tension and other signs of glaucoma, and remains liable to relapses of acute symptoms for varying periods, but in many "absolute" cases, especially those which follow the acute forms of glaucoma, changes occur sooner or later, leading to staphylomata, cataract, atrophy of iris, and finally to softening and shrinking of the globe.

"*Glaucoma fulminans*" means extremely severe and acute glaucoma, in which sight is abolished in a few hours.

As a rule glaucoma runs the same course in the second eye as in the first, but sometimes it will be chronic in one and acute or subacute in the other.

Explanation of the symptoms.—The increased tightness of the eye-capsule impedes the flow of arterial blood to, and of venous blood from, the retina, and thus lowers its functional activity. When the retinal vessels can be seen in glaucoma, the arteries are somewhat narrowed, and often exhibit spontaneous pulsation, whilst the veins are always engorged, though much less so than in neuritis. This deprivation of blood will first affect the peripheral parts, because more resistance has to be overcome in reaching them, and this probably explains the contraction of the visual field. The nutrition of the inner retinal layers suffers if the pressure is kept up (1) from the in-

sufficiency of arterial blood, and the changes, including hæmorrhage, which follow impeded venous outflow;
(2) from stretching and atrophy of the nerve-fibres on
the disc. The floor of the disc (*lamina cribrosa*),

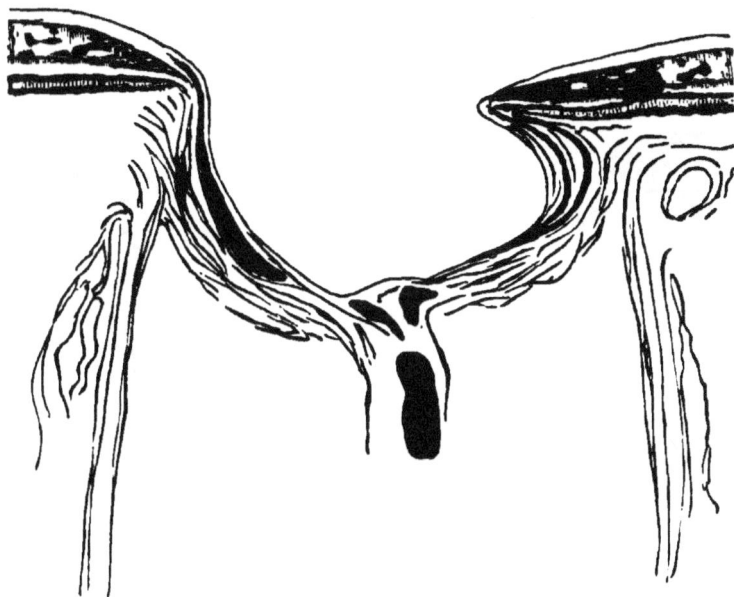

Fɪɢ. 44.—Section of very deep glaucoma cup. (Compare
Fig. 9.)

being the weakest part of the ocular capsule, slowly
yields in glaucoma, and is pressed backwards, the
nerve-fibres being proportionately dragged upon, displaced, and ultimately atrophied; the direct pressure on the nerve-fibres, as they bend over the
edge of the disc, also helps in the same process.
The result is that the disc becomes not only atrophied,
but depressed or hollowed out (Fig. 44). This hollow is the well-known " glaucomatous cup " which,
when deep, has an overhanging circumference, because
the border of the disc is smaller at the level of the
choroid than at the level of the *lamina cribrosa;*
even when shallow its sides are still quite steep
(Fig. 45).

With the ophthalmoscope this change is shown by a sudden bending of the vessels just within the border of the disc, where they look darker because foreshortened ; if the cup be deep they may disappear beneath its edge to reappear on its floor, where they have a lighter shade (Fig. 46.)

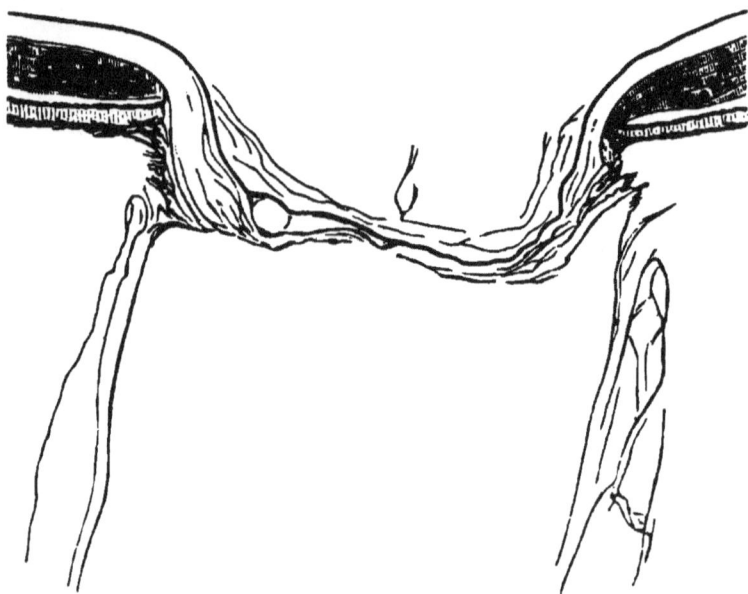

FIG. 45.—Section of less advanced glaucoma cup.

The vessels, as a rule, do not all bend with equal abruptness, some parts of the disc being more deeply hollowed than others, or some of the vessels spanning over the interval instead of hugging the wall of the cup. It is probable that several months of continuous increased tension are needed to produce cupping recognisable by the ophthalmoscope.

Although in many cases the excavation extends from the first over the whole surface of the disc, it is certain that this is not always so; the depression starts, in some of the most chronic cases, at the part naturally thinnest (the physiological pit or cup), and enlarges towards the periphery (p. 200). A deep cup is sometimes partly filled up by fibrous

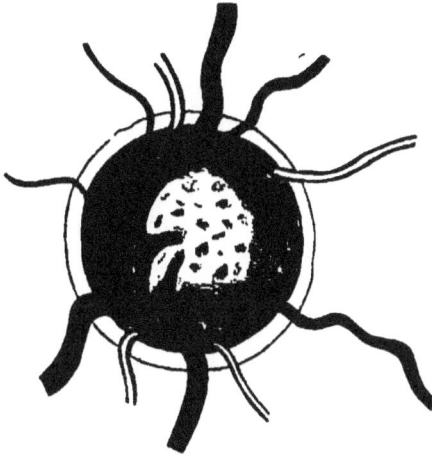

FIG. 46. Ophthalmoscopic appearance of deep cupping of the disc in glaucoma (altered from Liebreich).

tissue, the result of chronic inflammation, its true dimensions not being then appreciable by the ophthalmoscope.

The shallowness of the anterior chamber is due to the protrusion of the lens; it is not a constant symptom. The pressure on the ciliary nerves accounts for the somewhat dilated and immoveable pupil and for the anæsthesia of the cornea. In old-standing cases the iris is atrophied and discoloured, and may be shrunken to a narrow rim, but in uncomplicated glaucoma iritic adhesions are never seen. The corneal changes depend at first on mere "steaminess" of the epithelium, but in old cases the corneal tissue may be hazy. In recent cases the aqueous humour is somewhat turbid. The lens appears to lose some transparency even in fresh cases, if severe; in old cases, as already stated, it often becomes slowly opalescent, and finally quite opaque. It is generally stated that the vitreous humour becomes hazy during the attacks, especially in severe cases, but since it is just in these very cases that the cornea and aqueous are most dull, the statements about the vitreous are conjectural (p. 197). The internal pressure tends in acute cases to make the globe spherical, by reducing the curvature of the cornea to that of the sclerotic; it also in all cases weakens the accommodation by pressing on the ciliary nerves, and perhaps by causing atrophy of the ciliary muscle and preternatural flattening of the lens; these facts together explain the rapid decrease of refractive power (*i.e.*, rapid onset or increase of presbyopia) which is sometimes noticed by the patient (see p. 199).

The choroidal circulation is obstructed by the increase of pressure, whilst, on the other hand, its primary disturbance is believed to be often the starting-point of that increase, and in the following manner. The greater part of the venous blood of the choroid, ciliary body, and iris, leaves the eye by a few large trunks (*venæ vorticosæ*, Fig. 1), which pass very obliquely through the sclerotic at the equator; only a small part passes out through the numerous small twigs which perforate less obliquely near the cornea (*anterior ciliary veins* forming the episcleral plexus, p. 13). Increased tension once started will tend to close the oblique channels of the *venæ vorticosæ*, and this retention of blood will increase the pressure still more, a "vicious circle" being thus established. It can readily be understood that thrombosis or disease of these vessels might itself be the first factor in the obstruction. Obstruction, whether primary or secondary, of the *venæ vorticosæ* throws the blood into the anterior ciliary veins, the enlargement of which accounts for much of the dusky reticulated congestion of the ciliary region spoken of at p. 201. This imperfect collateral circulation, however, probalby could not of itself restore normal tension, even if obstruction to the escape of venous blood from the choroid were the only cause for the rise of pressure.

Etiology.—The local cause of the increased tension might conceivably be—(1) active contraction of the sclerotic, but there is no evidence that this occurs; (2), excess of the contents of the eye from increased supply of fluids; (3), the same from defective removal; and each of the latter two causes would be aided by the loss of distensibility which the sclerotic always undergoes with advancing age. There are differences in the thickness and distensibility of the sclerotic in different persons, and it is thicker in hypermetropic eyes, which are known to be particularly subject to glaucoma, than in myopic eyes. Many facts appear to tell in favour of the last hypothesis, that of defective removal. Recent pathological work has shown that changes are generally present near the attachment of the iris, ciliary muscle, and Schlemm's canal, which may with much probability be assumed to impede the escape of fluid from the anterior chamber, and perhaps from the vitreous. The openings in the fenestrated tissue of the *ligament. pectinatum iridis* are found to be obliterated either by adhesion between the iris and the cornea at their receiving angle, or by inflammatory exudation in the tissues around Schlemm's canal; and as it is by way of this angle between iris and cornea that much of the fluid transudes from the interior of the eye in health (whether into blood-vessels or lymphatics seems uncertain), the stoppage of this porous outlet would be expected to increase the tension. Slight increase of T. thus started might readily be kept up, and, by causing obstruction to venous outflow, be made the basis of severe acute attacks. The production of glaucoma by the retention of fluid in the posterior chamber, when iritis has left a circular synechia, seems also to support this view. Defective absorption by capillaries and veins probably causes increase of tension in certain morbid states of the blood particularly if, as is common in glaucoma, the small vessels are diseased. Obstructive disease of the coats of the venæ vorticosæ might, as already explained, be a cause; and that such disease does occur in some cases seems highly probable from the obvious thickening of the walls of choroidal vessels which is sometimes seen in old people (p. 160). Secondary glaucoma following wound of the lens with traumatic cataract is due to pressure on the periphery of the iris by the swollen lens. Glaucoma is very common in some cases of irido-cyclitis, especially the sympathetic form, when the parts above specified are the seat of structural changes, and in some of which there is believed to be also hypersecretion of aqueous.

Effect of over-supply of fluids.—Functional hyperæmia and active inflammation of the retina and choroid are not among the recognised causes of glaucoma. Dilatation of the arteries by vaso-motor paralysis is said to diminish the tension. Tumours in and even upon the eye often give rise to attacks of secondary glaucoma, and probably the initial factor in these is the occurrence of sudden attacks of congestion, or rapid increase of growth; certainly the glaucoma stands in no definite relation either to the size or position of the tumour. It is obvious that all these factors would be most potent if the sclerotic were too rigid.

A relation is observed in some cases between glaucoma and a previous, often a chronic liability to neuralgia of the fifth; and T. is said to be lowered in paralysis of this nerve. It is at present uncertain whether any affection of the fifth nerve has a direct influence upon then nutritive processes of the eye, or only acts indirectly by causing associated congestion, and thus setting up glaucoma in an eye predisposed to it.

General and diathetic causes.—In an eye predisposed by rigidity of sclerotic or by chronic changes in the tissues at the rim of the anterior chamber, or by disease of choroidal blood-vessels, any cause of congestion may precipitate an acute attack. A liability to congestion of the eyes in connection with digestive disturbances, gout or neuralgia, or the same result brought on by over-use of the eyes, especially of presbyopic and hypermetropic eyes when under-corrected by glasses, or a blow, or prolonged ophthal-moscopic examination, may all bring it about. Atro-pine, which has the power of increasing the eye-tension, has sometimes caused an attack, though other factors have been at work in some of the cases attributed to its influence. Iridectomy in one eye occasionally has the effect of precipitating the disease in the other, but its mode of action is unexplained. Glaucoma is commoner in women than in men, and after than before the age of forty-five. It is very rare in young adults and children, and is then gene-rally chronic and often gives rise to or is associated with other changes in the eyes. Acute cases are often dated from a period of overwork of the eyes, or of want of sleep, as from sitting up nursing, &c. There is often a history of gout and of neuralgia of the fifth nerve. Hence, patients who have had glaucoma in one eye should be strongly warned as to the danger of over-using the eyes, of work-ing without proper glasses, and against dietetic errors.

Treatment. — Until quite lately no treatment except iridectomy or an equivalent operation was thought admissible. It is now known that eserine (the active principle of Calabar bean) used locally diminishes the tension in glaucoma, and some few cases of acute glaucoma have been permanently cured by its means alone. The severity of the pain in recent acute cases may be much relieved by leech-ing, warmth to the eye, and opium, with derivative treatment, such as purgation, and hot foot-baths. If to these we add eserine there is hope that a cure may be effected in certain cases without operation. The question, however, is as yet unsettled, and iridectomy must still be considered the orthodox treatment for all cases of remediable glaucoma.

Iridectomy cures glaucoma by permanently reducing the tension to the normal or nearly normal pitch, but its mode of action is not fully known. It is found, however, that to insure success: (1) the path of the incision must lie partly in the sclerotic instead of wholly in the cornea, being at about $\frac{1}{20}$th to $\frac{1}{12}$th inch from the apparent corneal border; (2) the wound must be large, allowing removal of about a fifth of the circumference of the iris; (3) the iris should be removed quite up to its ciliary attachment, a result attained better by first cutting one end of the loop of protruding iris, then tearing it off from its ciliary attachment quite up to the other end of the wound, and cutting the other end of the iris separately, than by cutting across the loop at one stroke (*see* Operations). In some cases of quite recent and acute glaucoma the mere evacuation of the aqueous by paracentesis of the anterior chamber has been followed by permanent cure, probably by allowing the balance of the choroidal circulation to be restored. But when the disease is of longer standing this proceeding only gives temporary relief.

A mere wound in the sclerotic, differing but little in position and extent from that made for iridectomy, is sufficient to relieve + T., and to cure many cases of glaucoma permanently, and this novel operation (*sclerotomy*) is practised extensively by a few operators. But even if the removal of a piece of iris should be shown to be seldom necessary, iridectomy (performed as above) will probably always remain the best operation. Sclerotomy is objectionable on the following practical grounds: (1st) from the mode of operation the exact position and extent of the incision are not perfectly under control; if too far forward and too short the incision is insufficient, if too far back and too long there is risk of wounding the ciliary processes and getting hæmorrhage into the vitreous; even shrinking of the operated eye and sympathetic mischief in the other have occurred; (2nd) because although the wound is in great part sub-conjunctival the iris may prolapse so far into it as to need removal, and the operation then becomes an iridectomy.

Still more recently an operation for trephining a circular piece out of the sclerotic, behind the ciliary region, has been introduced by Argyll Robertson, in preference to iridectomy, but appears open to grave objections. Moreover, since it is known that the periphery (ciliary part) of the iris often adheres more or less to the cornea in glaucoma, it is quite probable that, apart from any objections on the ground of danger, sclerotomy, and allied procedures may be really insufficient, and that removal of a sector of the iris is necessary in many cases.

But whichever operation be employed the formation of the scar of the operation wound in the sclerotic is certainly a most important factor in the operations for glaucoma. This scleral wound, whether in iridectomy or sclerotomy, heals, unlike a corneal wound, with the interposition of a thin layer of new tissue, not by the immediate cohesion of the cut edges; and the most probable theory of the mode of action of iridectomy and sclerotomy appears to be that which refers its effect to the interposition of a layer of tissue, more permeable to the eye fluids than the natural coats (*filtration-scar* theory). The scar may also act as a sort of safety valve by yielding to pressure; and its efficacy has also been attributed to the formation of new vessels which, by communicating between the outer and inner surfaces of the globe, are supposed to allow a freer escape of blood. The practical fact that an iridectomy for glaucoma, which heals rather slowly, is thought to be more favorable than one which heals immediately, *i. e.* without new tissue, and that a slight bulging of the scar is believed by some surgeons to be rather a good thing than otherwise, are probably expressions of the real value of the new tissue formed during somewhat slow healing.

Scleral iridectomy is to be done in all cases of acute and subacute glaucoma, where some sight already remains, and whether there be great pain or not, and even if all p. l. be abolished, provided this be only of a few days' duration (see Operations). Even if the eye be permanently quite blind, iridectomy or sclerotomy may be done for the relief of pain in preference to excision of the globe (see Secondary Glaucoma and Tumours).

In very chronic glaucoma, when well developed, the rule is less clear, for it is well known that the effect of operation in such cases is far less constant. As no other treatment is of use, and iridectomy is certainly often beneficial, the operation should, as a rule, be done in confirmed cases, the patient's judgment being allowed a fair weight in the decision. The same difficulty occurs in some of the so-called "premonitory attacks," which are really early transient attacks of slight glaucoma. When once it is clear that such attacks of temporary mistiness and rainbows are glaucomatous, and that they are getting more frequent, the operation should, as a rule, not be deferred. An exception is, however, to be made if the patient can be seen at short intervals; eserine should then have a fair trial before operation is resorted to. It is to be remembered that iridectomy done when sight is still nearly perfect may, by allowing light to pass through the margin of the lens, and thus to undergo "spherical aberration," cause an increase of the defect; and this, though not of necessity a contra-indication, must be carefully taken into account. The patient's prospect of life must also be allowed for in chronic glaucoma; if he be old and feeble, life may end before the disease has in its natural course caused blindness.

The prognosis after iridectomy is, in general terms,

better in proportion as the disease is acute and recent. If operated on within about a fortnight of the onset of acute symptoms, and provided there be at least good p. l. at the time of operation, sight is usually restored to the state in which it was at the onset. If the attack be recent nearly perfect sight will be restored; if an acute be engrafted on a chronic attack sight will be improved more or less; if the case be chronic it will be prevented from getting worse. The prognosis in acute cases, however, varies a good deal with the severity as well as the acuteness. In cases combining the maximum of acuteness and severity (*glaucoma fulminans*) the operation may be successful, even if for a day or two all p. l. has been abolished.

The full benefit of the operation is not seen for several weeks, though a marked immediate effect is produced in acute cases. A slight degree of + T. sometimes remains permanently after iridectomy in cases of old standing, and does not appear deleterious, provided it be very much less than before the operation; the eye tissues can in some degree adapt themselves to increased pressure.

A second iridectomy in the opposite direction, or a sclerotomy, should be done if the T., having been reduced to normal, or very slightly +, after the first operation, rises definitely, and is accompanied by a return of other symptoms; but two or three weeks should generally elapse, for slight waves of glaucomatous tension may occur during states of temporary congestion or irritation before the eye has fully recovered from the first operation, and such symptoms may generally be relieved by other means. Cases which relapse definitely or which progress steadily after the first operation are always unfavorable, and the second operation must not be very confidently expected to succeed. If after iridectomy in acute glaucoma the symptoms are not relieved for a time, or become even worse, some deep-seated disease is to be suspected, such as hæmorrhage into the retina, or from the choroid, or a tumour (*see* Secondary Glaucoma).

Other treatment.—If for any reason we are obliged to delay the iridectomy against our judgment, the other means mentioned at p. 209 should be prescribed, including eserine drops used many times a day, and, if possible, a paracentesis of the anterior chamber. Stimulants should be forbidden or much reduced. After iridectomy, and until the eye has settled down to a permanently quiet state, all causes likely to induce congestion of the eyes must be carefully avoided, such as use of the eyes, stooping and straining, prolonged ophthalmoscopic examination, and the use of atropine. We should be on the alert for the earliest symptoms in the second eye after operation on the first (see p. 209), and the use of eserine and a few leeches may occasionally be advisable as prophylactics.

In a few cases of very chronic or subacute character, where high increase of T. is present, iridectomy seems to aggravate, instead of arresting, the disease, not being followed by even temporary benefit, but, by persistence of + T., increased irritability, and still further deterioration of sight ("*glaucoma maligna*"). It is believed that the forward tilting of the lens, which sometimes follows iridectomy, may help to account for these symptoms.

Glaucoma may occur independently in cataractous eyes; and in eyes from which the lens has been extracted, with or without iridectomy.

Secondary glaucoma may also be acute or chronic according as it is a consequence of active disease or of sequelæ. Thus, chronic glaucoma is often a consequence of circular iritic synechia with bulging of the iris (p. 103), and various forms of chronic iridokeratitis and irido-cyclitis, especially the sympathetic form, are liable to be accompanied by it. It

may follow perforating ulceration of the cornea with
large anterior synechia. The eye often becomes
temporarily glaucomatous in the course of traumatic
cataract, especially in patients past middle life (p.
145). In none of these cases is there much danger
of mistaking secondary for idiopathic glaucoma.

But secondary glaucoma may result from various
deeper changes. When the lens is dislocated, either
behind or in front of the iris, it often sets up glaucoma, and sometimes of a very severe type, apparently by pressing on the ciliary processes or iris.
There is generally the history of a blow; and in
posterior dislocation, even if the edge of the displaced
lens cannot be seen, the iris is usually tremulous and
its surface often bulging at one part and concave or
flat at another. If we are sure that a dislocated lens
is causing the symptoms it should be extracted by
a spoon operation (*see* Operations for Cataract),
and if lying in the anterior chamber should
always be removed. But in the glaucomatous state
of the eye after a severe blow it may be impossible
to feel sure of the condition of the lens, and then an
iridectomy must be done and the eye be watched ;
vitreous is very likely to escape at the operation if
there be dislocation of the lens, for the latter condition
implies rupture of the suspensory ligament. Hæmorrhage into an eye whose retina is detached (*e. g.* in
high degrees of myopia) may give rise to acute glaucoma with severe pain. A glaucomatous attack
generally occurs during the growth of an intraocular
tumour (p. 244). There will often be nothing in the
appearance of such an eye to distinguish the case from
an idiopathic glaucoma of the same severity and of
long standing, for, even if the lens be not opaque, and it
often is so, the other media will probably be too hazy
to allow an ophthalmoscopic examination. In almost
every case, however, the eye will be quite blind, and
will be known to have been so for weeks or months,
and there will also be the negative fact that the

fellow-eye shows no signs of glaucoma. A glaucomatous eye, which having been absolutely blind for
several months, remains painful and inflamed, and
the media of which are too opaque for ophthalmoscopic examination, should usually be excised as likely
to contain a tumour ; and especially if there be no
premonitory signs of glaucoma in the other eye.
Tumours in the eyes of children may cause secondary
glaucoma, but in these cases there is seldom any
difficulty in assigning the glaucoma to its right cause.
Secondary glaucoma now and then supervenes in cases
of albuminuric retinitis, and of embolism of the retinal
artery, and more commonly in some forms of retinal
and choroidal hæmorrhage. In the last-named cases
the diagnosis can sometimes be completed only after
an unsuccessful iridectomy has shown that the case
is not a simple one (p. 213).

CHAPTER XVII

THE optic nerve is often diseased in some part of its course, either within the skull, in the orbit, or at its intraocular end. The change may begin in the nerve, or pass to it from surrounding parts in the orbit or the eyeball, or be the consequence of disease in a distant part of the brain.

The relation between disease of the optic nerve and ophthalmoscopic changes in its visible portion, the optic disc (papilla optica), as well as the degree of affection of sight, varies greatly according to the seat, nature, and duration of the disease. The appearance of the disc may be entirely altered by œdema and inflammation, without the nerve fibres losing their conductivity, and, therefore, without loss or even defect of sight; on the other hand, morbid changes of inflammatory or atrophic character, causing destruction of the nerve fibres, may take place in the nerve trunk and produce great defect of sight, with little or no immediate change in the appearance of the disc. The three factors—visible changes in the disc, affection of sight, and nature of the morbid process—are very differently related in different cases. Although we are here concerned chiefly with the two former, a few words of explanation are needed as to the third.

The pathological changes to which the optic nerve is liable include those which affect other nerve tissues. Inflammation varying in seat, cause, and rapidity, and resulting in recovery or atrophy, may originate in the nerve itself, or may pass down the nerve

from the brain (descending neuritis), or may extend
into it from parts around ; atrophy may occur from
pressure, either by tumours or by distension of neighbouring cavities (*e.g.* the third ventricle) ; and the
optic nerve is very liable to the change known as
"grey degeneration," "sclerosis," or "parenchymatous neuritis."

Lastly, the optic nerve being surrounded by a
lymphatic space (the "subvaginal space" of the
optic nerve), which is continuous, through the
optic foramen, with the meningeal spaces in the
skull, and is enclosed in a tough fibrous tube (the
"outer sheath" of the optic nerve), is liable to compression by fluid, and by inflammatory products formed
in that space. Such accumulation, either by retention or secretion, in the subvaginal space is
believed by many observers to frequently furnish the
true explanation of the "optic neuritis" about to be
described, which is so commonly associated with intracranial disease. The nerve at its entrance into
the eye is enclosed by a non-distensible collar of
sclerotic, and any compression of the nerve just behind the eye, as by fluid collected in the sheath, will
proportionately retard the exit of venous blood. A
slight degree of œdema of the optic disc so produced
will tend to increase itself, so long as the compression
continues, by pressing the veins as they leave the eye
against the unyielding scleral ring of the disc; and
if the process be intense and last long enough,
secondary inflammation, degenerative changes, and
atrophy, may come on in the disc. The early part
of the process may be compared to the effect of a
string tied round the finger. The trunk of the nerve
is believed to be healthy in very many of these cases.
But our knowledge of the changes in the optic nerve,
in optic neuritis, and of their relation to the various
intracranial diseases, which so often cause it, is as yet
very incomplete, and some observers, whose opinions
are entitled to great weight, believe that true de-

scending neuritis is much commoner than has been supposed, and that it supplies the explanation of many cases which have been attributed to compression by the distended sheath.*

Inflammation may extend into the disc from the retina or choroid near to it, and it may occur in consequence of the sudden arrest of the blood-current in embolism and thrombosis of the central retinal vessels in their course through the nerve (pp. 161, 170, *et seq.*).

Ophthalmoscopic appearances of inflammation of the optic disc.—The changes caused by mere œdema of the disc are so mixed up with those of congestion and inflammation that it is for clinical purposes useless to attempt any sharp distinction between "swollen disc," "choked disc," or "ischæmia papillæ," attributed to strangulation of the disc at the scleral ring, and "optic neuritis" caused by inflammation descending the trunk or sheath of the nerve. The terms "neuritis," "optic neuritis," and "papillitis,"† will be used indifferently for these conditions.

The earliest changes in ordinary optic neuritis (papillitis) are reddening of the disc, blurring of its border by a greyish opalescent haze, distension of the large retinal veins, and swelling of the disc above the surrounding retina, as shown by the abrupt bending of the vessels just outside its margin. The patient may die or the disease may recede at this stage. But generally more decided changes occur ; the haziness becomes decided opacity, more or less obscuring the central vessels, covering and extending beyond

* For a full and masterly statement of this difficult subject, enriched with many new facts, the reader is referred to Dr Gowers' recent ' Manual and Atlas of Medical Ophthalmoscopy' (p. 63).

† "Papillitis " has been proposed by Leber as designating the ophthalmoscopic appearances of the inflamed or swollen optic disc, without implying any theory of causation.

the border (Fig. 47), so that the disc appears considerably increased in diameter; its colour becomes a mixture of yellow, and pink with grey or white, and it looks striated or fibrous, appearances due to a whitish opacity of the nerve fibres mingled with numerous small blood-vessels and hæmorrhages.

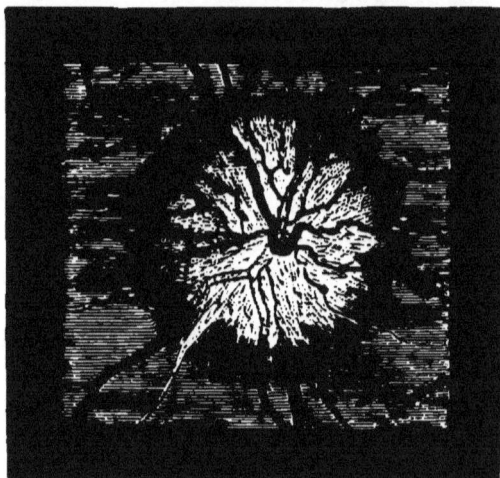

FIG. 47—Ophthalmoscopic appearance of severe papillitis. Several elongated patches of hæmorrhage near border of disc. (After Hughlings Jackson.) Compare with Fig. 48.

The veins often become larger and more tortuous, even kinked or knuckled; the arteries are either normal or somewhat contracted; there may be blood patches. Further increase in the swelling of the disc is shown by sudden bending and foreshortening of the large retinal vessels at the border of the hazy area, the foreshortening being proved by the darker colour and altered surface reflection at the bends. The prominence is realised chiefly by attention to the above changes in the course of the vessels, for with the monocular ophthalmoscope the fundus is not seen stereoscopically.*

* For good illustrations of the appearances in different stages see Gowers, Pl. I, 5 and 6, and Pl. II, 1.

Such changes may disappear, leaving scarcely a
trace; or a certain degree of atrophic paleness of the

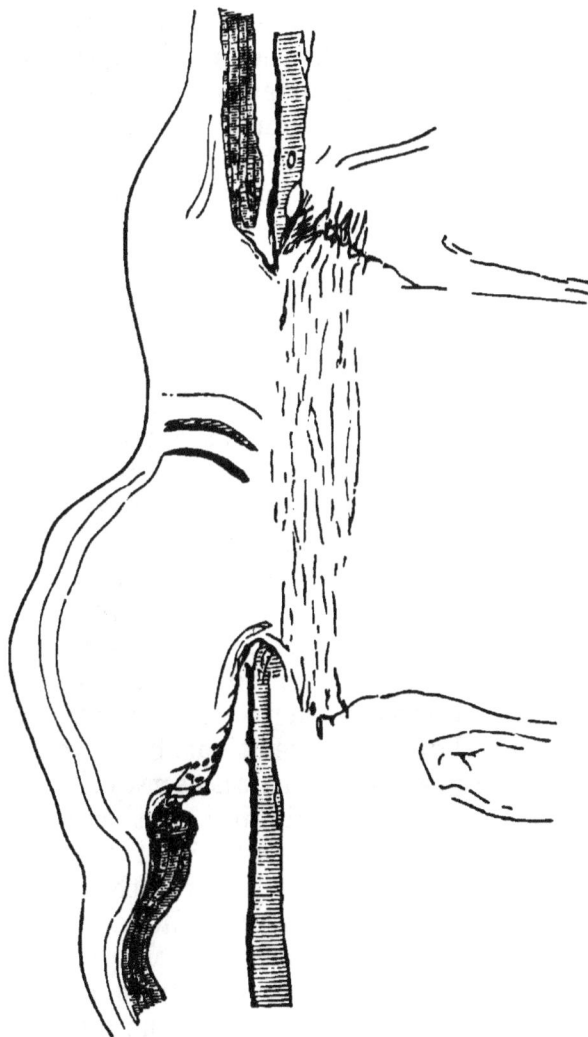

FIG. 48. Section of the swollen disc in papillitis, showing that the swelling is limited to the layer of nerve-fibres (unshaded). The other (shaded) retinal layers are not altered in thickness. (Compare with Fig. 9.)

disc, with some narrowing of the retinal vessels and
thickening of their sheaths, or other slight changes,
may remain. But in many cases the disc gradually,
in the course of weeks or months, passes into a state
of atrophy; the opacity first becomes whiter and

smoother looking ("woolly disc"); then it slowly clears off, generally first at the side next the yellow spot, and the retinal vessels simultaneously shrink to a smaller size, though they often remain tortuous at the disc for a long time. As the mist lifts, the sharp edge, and finally the whole surface of the disc, now of a staring-white colour, again comes into view. A slight haziness often remains, and the boundary of the disc is often notched and irregular; but these are not signs upon which too much reliance must be placed. The degree to which the central vessels are shrunken is one of the best signs of the degree of atrophy of the nerve.

Sight is seldom much affected until marked papillitis has existed some little time; if the morbid process quickly ceases no failure may take place, or sight may fail, may even sink almost to blindness for a short time, and more or less complete recovery take place if the changes cease before compression of the nerve-fibres has given rise to atrophy. Gradual failure late in the case, when retrogressive changes are already visible at the disc, is a bad sign. The sight does not change, either for better or worse, after the ophthalmoscopic signs of papillitis have passed off, and though the relations between sight and final ophthalmoscopic appearances vary, it is true in general (1) that great shrinking of the principal retinal vessels indicates great defect of sight, and generally accompanies extreme pallor with some permanent residual haziness of the disc (advanced postneuritic atrophy); (2) that considerable pallor, and other slight changes, such as white lines bounding the vessels and caused by increase of the interstitial connective tissue of the disc, after papillitis, are compatible with fairly good sight, if the central vessels are not much shrunken.

White atrophy, with blindness following undoubted papillitis, does not always show signs of the past violent inflammation; the appearances may indeed be indistinguishable from those caused by primary atrophy.

Papillitis is double in the great majority of cases; occasionally one-sided only, and then generally indicative of disease in the orbit. In the double cases however, there are often inequalities, in time, in degree, and in final result, between the two eyes.

Etiology.—Besides the two explanations of papillitis already offered (p. 218, descending neuritis and distension of the nerve sheath), the process has been attributed to compression of the ophthalmic vein in the cavernous sinus, as the result of increased intracranial pressure, and to some change in the vasomotor nerves of the optic nerve vessels, caused by the irritation of the intracranial disease. The former theory is quite inadequate, unless in a few cases where the nerve is compressed in the orbit, and the latter is entirely hypothetical. What is now most wanted is careful microscopical examination of the optic nerve in its whole length in the early stages of papillitis. At present it can only be said that the sheath is often found distended at the post mortem in cases of papillitis; and that papillitis occurs chiefly in cases of irritative intracranial disease, viz. in meningitis, both acute and chronic, and in intracranial new growths of all kinds, whether inflammatory (syphilitic gummata), or neoplastic; and that the chiasma and optic tracts, although often appearing, macroscopically at least, quite healthy, in other cases are obviously inflamed or softened. It is rare in cases where there is neither inflammation nor tissue growth, as in cerebral hæmorrhage and intracranial aneurism. Further, it must be stated that no constant relation has been proved between papillitis and the seat or extent of the intracranial disease. Papillitis has occasionally been found without coarse disease, but with widely diffused minute changes, in the brain.

Thus, the occurrence of papillitis, although

pointing very strongly to organic disease within the skull, and especially to intracranial tumour, is not of itself either a localising or a differentiating symptom. It is probable that in some cases both distension of the optic sheath behind the eye and descending inflammation of the optic nerve within the skull, may co-operate. Localised intracranial disease, e. g. thrombosis of the cavernous sinus or inflammation about the sphenoidal fissure, as well as tumours and inflammations in the orbit, are occasional causes of papillitis and of descending neuritis, which may then be one-sided. In some of these there will be protrusion of the eye and affection of other orbital nerves, and the exact seat of disease may be very difficult to diagnose.

The eye changes are not always limited so strictly to the disc and its border (pure papillitis), for in some cases a wide space of surrounding retina is hazy and swollen, exhibiting hæmorrhages, white-plaques, or even lustrous white dots at the yellow spot, very like what is seen in renal retinitis (papillo-retinitis, neuro-retinitis, p. 167). It is not always easy, if possible, to distinguish between such changes and renal retinitis, especially as in the latter there is sometimes great swelling of the disc (p. 179). In renal cases there is always albuminuria, the patient is seldom a young child, and it is said that the cases with most neuritis occur in an advanced stage of the kidney disease ;* in the cases of neuro-retinitis, most closely resembling renal cases, but caused by cerebral disease, there will be no albumen, and the changes will seldom closely resemble those of albuminuria until they have existed for long and caused very great defect of sight.

In a few cases well-marked double papillitis occurs without other symptoms and without assignable

* Gowers.

cause. Other occasional causes of double papillitis, with or without retinitis, are lead poisoning, the various exanthemata, sudden suppression of menstruation, and, perhaps, exposure to cold.

Certain cases of failure of sight, often in only one eye, with slight neuritic changes at the disc, followed by recovery or by atrophy, are probably to be referred to chronic neuritis behind the eye (*retro-bulbar neuritis*). The changes are clinically very different from any of those above described. Some of these cases are, perhaps, comparable with the so-called rheumatic paralysis, *e.g.* of the facial nerve; others are accompanied by severe headache and other symptoms of intracranial disease.

Syphilitic disease within the skull is a common cause of papillitis, but the eye changes alone furnish no clue to the cause, which may operate as follows: (1) by giving rise to intracranial gumma not in connection with the optic nerves, but acting as any other tumour acts; (2) by implication of the chiasma or optic tracts in gummatous inflammation, which causes descending neuritis; (3) in rare cases neuritis ending in atrophy and blindness occurs, in secondary syphilis, with severe head symptoms pointing to acute meningitis.

The condition of the pupil in neuritic affections depends partly on the state of sight and partly on the rapidity of failure. As a rule, in amaurosis from atrophy of the discs after papillitis the pupils are for a time rather widely dilated and motionless; after a time they often become smaller, and regain a certain amount of mobility to light.

ATROPHY OF THE OPTIC DISC

By this is meant atrophy of the nerve-fibres of the disc, and of the capillary vessels which feed it. It is shown by change of colour, and in most cases by a præternatural sharpness of outline. The central retinal vessels may or may not be shrunken. The

15

disc is too white; milk-white, bluish, greyish, or yellowish in different cases. Its colour may be quite uniform, dead or opaque looking; or some one part may be whiter than another; the stippling of the *lamina cribrosa* (p. 33) may be more visible than in health, or, on the other hand, entirely absent, as if covered or filled up by white paint. The choroidal boundary is too sharply defined; it may be even and circular, or irregular and notched. Within it the sclerotic ring (p. 33) is often seen with unnatural clearness as a ring of even whiter colour than the nerve which it encircles. Mere pallor of the disc, as is present in extreme general anæmia, must not be mistaken for atrophy; the change is then one of colour only; there is neither morbid distinctness of parts, nor disturbance of outline. The large retinal vessels are to be carefully noted as to size and tortuosity, both of which points are important for diagnosis of cause and for prognosis.

Local causes. — (1). The nerve-fibres undergo atrophy in the process of absorption and shrinkage of newly-formed connective tissue, after severe neuritis affecting either the disc alone or the whole length of the nerve (see p. 220). This is known as "atrophy after neuritis," "consecutive" or "post-papillitic" atrophy.

(2.) When the disc participates secondarily in inflammation of the retina or choroid it also participates in the succeeding atrophy (pp. 161, 170, &c.)

(3.) Atrophy of any part of the optic nerve trunk or tract, whether from pressure, as by a tumour or by distension of the third ventricle in hydrocephalus, from injury, or localised inflammation, leads to secondary atrophy, which sooner or later becomes evident at the disc. Such cases often show the conditions of pure atrophy, without complication, either by adventitious opacity or disturbance of outline, and often without change in the retinal vessels. They are not common.

(4.) The optic nerves are liable to chronic sclerotic changes, chronic inflammation of neuroglia ("parenchymatous neuritis "), and of the connective-tissue framework ("interstitial neuritis,") ending in atrophy without the occurrence of papillitis. The change in these cases appears to begin at the disc, but the exact order of events is not fully known in this large and important group.   One or other of these changes is the pathological equivalent of the appearances known clinically as " primary " or " progressive" atrophy.

Clinical aspects of atrophy of the discs.—As in optic neuritis, so in atrophy and pallor of the disc, there is no invariable relation between the appearance (especially the colour) of the disc and the patient's sight. A considerable degree of pallor, which it may be impossible to distinguish from true atrophy, is sometimes seen with excellent central vision (p. 18), though usually accompanied by some defect of the visual field. Again, it is often the case that the discs will look just alike, although the sight is much better in one eye than the other.

Patients with atrophy of the disc come to us because they cannot see well or are completely amaurotic. There are usually no other local symptoms except such as are furnished by the pupils, and in this respect cases of double optic atrophy present many variations.   In post-papillitic atrophy the pupils are generally partly dilated and sluggish or motionless; in most cases of primary progressive atrophy they are of ordinary size, or actually smaller than usual, and act very imperfectly.   When only one eye is affected, the other being quite healthy, the pupil of the amaurotic eye may be a little larger than its fellow, and is found to act only in association with it (p. 14).

The visual field, in cases of atrophy, is generally contracted, or shows irregular invasions or sector-like defects. Colour-blindness is a marked symptom in nearly all cases of atrophy, but is not always proportionate to the loss of vision, being in some much greater and in others much less than the acuteness of vision would lead us to expect (*see* also Amblyopia).

Green is the colour lost soonest in nearly all cases, and red next, but in this respect variations are occasionally observed.

A. Cases in which both eyes are affected may be conveniently classified as follows in regard to diagnosis and prognosis.

(1.) If the atrophic changes point decidedly to previous papillitis (p. 221) we may say that sight will certainly not improve; if the case is recent it may for a time get worse. The case must, of course, be investigated most carefully as to the cause of the neuritis.

(2.) Whenever, with atrophy of the disc, the retinal arteries are much shrunken, whether neuritis or retinitis have occurred or not, the prognosis is bad (p. 222).

(3.) The most careful examination leaves it uncertain whether previous papillitis have occurred. Still, as consecutive cannot always be distinguished from primary atrophy (p. 222), inquiry should be made for previous symptoms of intracranial disease. But in a large number of the cases, which present no ophthalmoscopic evidences of previous papillitis, the history will also be quite negative as to cerebral symptoms; and these will, for the most part, fall into the two following groups.

(4.) Cases in which there are symptoms of chronic disease of the spinal cord, usually of locomotor ataxy; or, much more rarely, symptoms of general paralysis.

(5.) Cases in which no spinal symptoms can be made out and no cause assigned for the atrophy; these form a majority of all the cases of progressive atrophy that come under care.

The sclerosis leading to atrophy of the discs in locomotor ataxy has been held to come on at a comparatively late stage of the disease, when well-marked spinal symptoms have appeared. In such cases the optic atrophy always becomes symmetrical, though it generally begins some months sooner in one eye than in the other; it always progresses, though sometimes very slowly, to complete or all but absolute blindness; and the discs are usually characterised by a uniformly opaque, grey-white colour, the lamina cribrosa being concealed, although neither the central vessels nor the disc margin are obscured in the least. The central vessels are often not materially lessened in size, even when the patient is quite blind.

Numerous cases of progressive atrophy are seen which agree in every respect with the above, but where no signs of spinal-cord disease are present, even though the patient has been long blind. It is now known that in many of these patients ataxic symptoms come on sooner or later, and it is in the highest degree probable that could the cases be followed up for a sufficient number of years this result would be found to be the rule. Indeed, preataxic optic atrophy is now a recognised method of onset of the disease ; and though our information is incomplete, and we do not yet know in what proportion of cases of optic atrophy the eye-disease remains uncomplicated, enough is known to justify the assertion that the optic atrophy of ataxy is very often an early, and seldom a late, symptom of the disease. Cases of this class (5) are far commoner in men than in women.

In making the prognosis of cases of slowly progressive uncomplicated amblyopia or amaurosis, with more or less atrophy of discs, special attention is to be paid to whether or not the failure was synchronous, and whether it is now equal in the two eyes (compare p. 232). The examination of the field of vision in early cases is also of much importance, though more difficult to make ; peripheral contraction, as distinguished from central defect, is a bad sign. Cases of gradual uncomplicated failure of sight, in which the symptoms have from the beginning been equally symmetrical, will generally be found to show but slight atrophic changes in proportion to the defect of sight, and their consideration is best deferred to the subject of amblyopia.

B. Single amaurosis with atrophy of the disc, in a majority of cases indicates former embolism of the central artery (p. 181), or some transient local affection of the optic nerve (pp. 224 and 231). The latter cases are often accompanied by severe, localised headache or neuralgia. But in cases of progressive atrophy, accompanying or preceding spinal disease, a very long interval occasionally separates the onset of the disease in the two eyes, and we may see the first before the commencement of disease in the second.

AMBLYOPIA

The term means dulness of sight, but its use is generally restricted to cases of defective sight, short of blindness, in which the visible changes are disproportionately slight. The word amaurosis is used for more advanced cases of the same kind, complete blindness without apparent cause. These terms are essentially clinical ones, as distinguished from papillitis and atrophy, which imply easily recognised pathological changes in the disc. The amblyopia may depend upon disease in the retina, or in any part of the optic nerve or tract, or in the optic centres ; and it may be a temporary or a permanent condition. In the investigation it is always most important to distinguish single from symmetrical cases.

Two groups of unsymmetrical amblyopia may be considered first, because they are common, and negatively important.

(1.) *Amblyopia from suppression of the image* in one eye in cases of squint. A squinting person, in order to avoid the difficulties attending double vision, suppresses the consciousness of the image formed in the squinting eye. If this suppression is continued the sensorium permanently loses the power of perceiving images in this eye, and we say that the eye becomes amblyopic when we ought to say that the corresponding centre loses perception. It is seldom worth while to treat this defect, although oft-repeated separate practice of the squinting eye, the sound eye being closed, leads to some improvement of sight. Sight generally remains best in the temporal part of the field in this amblyopia. The suppression is much more easily effected by some persons than others, and early in life than in later years; hence, children who have squinted constantly since an early age often do not have diplopia, while in adults diplopia often continues for a very long time. When the suppression is temporary, even though often repeated, as in cases of alternating and of periodic squint in hypermetropia, no harm results.

(2.) Amblyopia from defective retinal images. In cases of high hypermetropia or astigmatism, when clear images have never been formed, the correction of the optical defect by glasses at the earliest practicable age fails in a large number of cases, at any rate for a time, to give full acuteness of sight. The defect is presumably due chiefly to want of education in the appreciation of clear images. Probably the same explanation applies to the numerous cases in which, with anisometropia, the sight of the more ametropic (p. 249) eye even when corrected by the proper glasses remains defective, although no squint have ever existed; and in some degree at least to the defect so often observed after perfectly successful operations for infantile and lamellar cataract. When discovered late in life such defect is seldom altered by correcting the optical error, but in children the sight improves when the suitable glasses are constantly worn.

In cases of amblyopia not belonging to either of these categories a definite history of onset can generally be elicited. Two principal divisions may be

formed according as the defect is single or double.
In regard to the former we must remember that
it is very common for defect or even total blindness
of one eye to remain unnoticed for a long time, even
many years, until accidentally discovered by closing
the good eye. In such cases the patient is often
alarmed and naturally believes, until undeceived,
that the failure is recent.

(3.) Cases of recent failure of one eye with little
or no visible change occur, but are rare. They are
generally seen in young adults; the onset is often
rather rapid, and accompanied by neuralgic pains in
the same side of the head. There may be pain in
moving the eye or tenderness when it is pressed back
into the orbit. Sight may be very defective, especially
at the centre of the field. There are usually slight
differences between the two discs, that of the affected
eye being often hazy and congested. The attack is
often attributed to exposure to cold. The majority of
these cases recover under the use of blisters and
iodide of potassium, but I believe that in a certain
number the defect is permanent, and that the disc
becomes atrophied. A slight and transient peripheral neuritis near the eye supplies the most likely
explanation, and these cases may perhaps be analogous to peripheral paralysis of the facial nerve
(p. 224).

(4.) Much commoner is a progressive and equal
failure of sight in both eyes, amounting in a few weeks
or months to great defect (14, 16, or 20 Jaeger),
with no other local symptoms except perhaps a little
frontal headache, but often with general want of
tone, nervousness, and loss of sleep and appetite.
The ophthalmoscopic changes are never pronounced,
and sometimes quite absent. At an early period
the disc is often decidedly congested, and, together
with the surrounding retina, slightly swollen and
hazy, but all these changes are so ill-marked that
competent observers may give different accounts of

the same case. Later the side of the disc next the yellow spot becomes pale, and later still, in bad cases, the pallor extends to the whole papilla, and the diagnosis of incomplete atrophy is given. The defect of sight is usually greatest in bright light, being less apparent early in the morning and towards evening. The pupils are normal in all respects, or at most rather sluggish. The defect is limited to, or at least is greatest at, the central part of the field (causing a central scotoma), or more strictly occupies an oval patch extending from the fixation point (corresponding to the yellow spot) · outwards to the blind spot (corresponding to the optic disc), and there is always a correspondingly defective perception of the central colours, green and red. These points may be detected accurately by the perimeter, and, indeed, in many cases no colour defect is apparent if the patient be tested with large masses of colour. Often the patient will, by his statements, give us a clue to the character of the defect by saying that he mistakes sovereigns for shillings, that he cannot see the middle of the face of the person he looks at, or that people's faces look pale. The periphery of the field being good, no difficulty is experienced in going about, the large surrounding objects being visible; hence the patient's manner differs from that of one with progressive atrophy, who finds difficulty in walking about, &c., because his visual field is contracted at the periphery (p. 229).

The patients are, almost without exception, males, and almost always at or beyond middle life. With very rare exceptions they are all smokers, and have smoked for many years, and a large number are also intemperate in alcohol.

The exceptions occur chiefly in a very few patients in whom a similar kind of amblyopia is hereditary, is liable to affect the female as well as the male members, and sometimes comes on much earlier in life. The causation of this disease is obscure.

In the common cases all are agreed that tobacco has a large share in the causation, and in the opinion of an increasing number of observers it is the sole exciting cause. The share to be attributed directly to alcohol, and to the various causes of general exhaustion, such as anxiety, underfeeding, and general dissipation, cannot be considered settled so long as many competent observers hold that the disease may be brought on by such influences collectively without the influence of tobacco. My own opinion, based on the examination of a considerable number of cases, is that tobacco is the essential agent, and that the disuse or diminished use of tobacco is the one essential measure of treatment. It is important to remember that the disease may come on either when the quantity or strength of the tobacco is increased, or when the general health fails and a quantity which was formerly well borne becomes excessive. Hence, excepting the rare form above mentioned, the cases in this group may correctly be named " Tobacco Amblyopia."

The prognosis is good if the case come to treatment early, and if the failure have been comparatively quick. In such cases really perfect recovery may occur; and an improvement so striking that the patient considers his recovery perfect is the rule. In the more chronic cases, or cases where already the whole disc is pale, a moderate improvement, or even an arrest of progress, is all we can expect. If smoking be persisted in no improvement takes place, and the amblyopia increases up to a certain point, but complete blindness very seldom if ever occurs. In the treatment disuse of tobacco is the one essential. If the man drinks too much, he should, of course, lessen the amount. It is customary to administer moderate doses of nux vomica, or to inject strychnia subcutaneously for a considerable period, but whether any medicine acts otherwise than indirectly, by improving the general tone, is quite doubtful. The pathology of the disease is obscure, but it is believed to depend on a mild and chronic inflammatory affection of the ocular end of the optic nerve.

Cases of double and equal amblyopia with little or no ophthalmoscopic change are occasionally seen, in which there is no reason to suspect tobacco, alcohol, intracranial disease, nor inheritance. These are always obscure and difficult cases.

Malingering. — Patients now and then pretend defect or blindness of one or both eyes, or exaggerate an existing defect; or, again, sometimes secretly use atropine to paralyse the accommodation. In most cases the imposture is evident from other circumstances, but sometimes great difficulty is found in detecting it.

The pretended defect of sight is usually confined to one eye. If the patient is in reality using both eyes, a prism held before one (by preference the "blind" one) will produce double vision; a prism of 6° or 8°, held base upwards or downwards, is best (p. 23). Another test, when only moderate defect is asserted, is to try the eye with various weak glasses, and note whether the replies are consistent; very probably a flat glass or a weak concave will be said to "improve" or "magnify" very much. Again, atropine may be put into the *sound* eye, and when it has fully acted the patient be asked to read small print with both eyes; if he reads easily the imposture is clear, for he must be reading with the so-called "blind" eye. If absolute blindness of one be asserted the state of the pupil will be of much help (unless the patient have used atropine); if it acts independently of its fellow, and freely, the retina and nerve cannot be very defective (p. 14).

Asserted defect of both eyes is more difficult to expose, and indeed, it may be impossible to absolutely convict the patient if he is intelligent and has had access to means of information. The state of the pupils, of the visual fields, and of colour perception, are the most reliable tests.

CHAPTER XVIII.

TUMOURS AND NEW GROWTHS

A. For tumours and growths of the eyelids see pp. 46—49.

The following may here be added.

Nævus may occur on the eyelids, and implicate the conjunctiva, both of the lids and eyeball. Deep nævi may degenerate and become partly cystic.

Dermoid tumours (cystic) are not uncommon, being generally situated at the outer end of the eyebrow, much more rarely near the inner canthus. They lie beneath the orbicularis, the subjacent bone being sometimes superficially hollowed, and differ from sebaceous cysts by their greater depth and by the absence of adhesion to the skin. They often grow faster than the surrounding parts, and may then need extirpation, the thin cyst wall being carefully and completely removed through an incision parallel with and situated in the eyebrow. They contain, besides sebaceous matter, some short hairs.

A small *marginal fleshy wart* is sometimes seen on the free border of the lid. It is red, moist, and attached by a broad pedicle, and moulded both against the border of the lid and against the cornea. It is not definitely papillary, and sometimes seems to arise from the orifice of a meibomian canal. It causes some irritation, and is best snipped off.

B. TUMOURS AND GROWTHS OF THE CONJUNCTIVA AND FRONT OF THE EYEBALL

Cauliflower warts, exactly like those on the glans penis, are sometimes seen on the ocular and palpebral,

conjunctiva and caruncle. They have narrow pe-
dicles, and the body of the wart is flattened by
pressure like a cock's-comb. They should be snipped
off, but fresh ones are apt to spring up.

Lupus of the conjunctiva is generally accompanied
by lupus of the skin, and sometimes of the oral mucous
membrane. The conjunctiva is thickened, irregularly
tubercular, and very vascular. The disease very seldom
attacks the ocular conjunctiva, and is usually confined to a part of one eyelid. It is much benefited by
repeated carefully localised cauterisations with pure
nitrate of silver or the actual cautery.

The eyelid is now and then the seat of diffused
gummatous inflammation in the tertiary stage of
syphilis. The tarsus is especially affected and its
infiltration gives rise to a hard, indolent swelling of
the whole lid.

Chancres and tertiary syphilitic ulcers on the lids
have been referred to at p. 49.

Pinguecula is a little spot of yellowish colour,
looking like adipose tissue in the conjunctiva, close
to the inner or outer edge of the cornea. It consists
of thickened conjunctiva and subconjunctival tissue,
and contains no fat. It is commonest in old people,
and in those whose eyes are much exposed to local
irritants. It is sometimes a cause of apprehension to
its possessor, but is not in reality of the slightest
consequence.

Pterygium is a triangular patch of thickened conjunctiva, generally placed in the palpebral fissure,
its apex pointing towards, or encroaching upon, the
cornea. Pterygia vary much in thickness, vascularity,
and size. The disease is to be distinguished from
opacity of the cornea following ulceration, and from
the cicatricial bands (symblepharon) which often
form between lid and globe after burns or wounds of
the conjunctiva. Pterygium is rare in English practice, being seldom seen except in those who have at
some time lived in hot countries. It is often

progressive. The best treatment is to dissect it up from its apex and transplant it into a cleft in the conjunctiva below the cornea; this is more effectual than excision or ligature.

Cysts of minute size with thin walls and clear watery contents are not uncommonly seen in the ocular conjunctiva near the inner or outer canthus; sometimes they are elongated and beaded. They are probably formed by distension of the valved lymphatic trunks running from the neighbourhood of the cornea towards the canthi.

Dermoid tumours (*solid*) of the eyeball are much scarcer than the cystic dermoids of the eyebrow (p. 236). They are whitish, smooth, hemispherical and firm, and generally placed in the palpebral fissure. They may be wholly on the conjunctiva and movable, or partly on the cornea and fixed. They are solid, and hairs may grow from their surface. They are often combined with other congenital anomalies of the eye or lids. When seated on the cornea they cannot be entirely removed.

In some cases of *episcleritis*, quite a considerable swelling may form, and be mistaken for a tumour (see p. 110).

Malignant tumours arise much less commonly on the front of the eye than in the choroid or retina. They may be either epithelial or sarcomatous. An injury is often stated to be the cause of the growth.

Epithelioma may begin on the ocular conjunctiva, in which case it remains movable, or at the sclero-corneal junction, when it quickly encroaches on the cornea, infiltrates its superficial layers and becomes fixed. It may be pigmented. When such a growth is not seen until late it will perhaps be as large as a walnut, cover, or surround the cornea, and present a papillary or lobulated surface, and the glands in front of the ear may be enlarged.

Sarcoma in this region may be either pale or pigmented. It generally arises at the sclero-corneal junction, and when small the conjunctiva is traceable over it. But in advanced cases it may be impossible from the clinical features, to diagnose the nature of a tumour in this part.

Movable tumours (epithelioma) not involving the cornea may be cut off, but are very likely to recur; and recurrence is still more likely in the case of growths fixed to the cornea or sclerotic. Removal of the eyeball at an early date, especially in the case of sarcomata, is the best course in the majority of cases.

The lachrymal sac is sometimes the seat of morbid growths, which may be at first mistaken for chronic thickened mucocele (p. 54).

Fibro-fatty growth, in the form of a yellowish, lobulated, or tongue-like protrusion from between the lid and the globe, is rather a curiosity than of much importance. It generally lies in front of the lachrymal gland. It is believed to be congenital, but is apt in after-life to grow more in proportion than the surrounding parts.

Cystic tumours may be met with beneath the palpebral conjunctiva. Some are caused by occlusion and distension of a duct of the lachrymal gland (p. 52), but others cannot be so explained (*see* Nævus). Fibrous, and even bony tumours are occasionally seen in the substance of the upper lid, perhaps starting from the tarsus; and soft, pedunculated (polypoid) growths have been met with in the sulcus between lid and globe.

C. TUMOURS IN THE ORBIT

Protrusion of the eye (*proptosis*), with or without lateral displacement and limitation of its movement, is always the result of a tumour of any notable size in the orbit. As a rule, there are no inflammatory symptoms (see exceptions below). It is obvious that the diagnosis of the size, attachments, and nature of growths in the orbit, must often be doubtful even for the most experienced surgeon, since the deep parts of the orbit cannot be explored.

A tumour in the orbit may have originated in one of the proper orbital tissues or in the lachrymal gland, in the periosteum, upon or within the eyeball, from the optic nerve; or it may have encroached upon the orbit from one of the neighbouring cavities. Tumours in the orbit when fluctuating may be either cystic or ill-defined; they may be pulsating or solid, and either movable or fixed by broad attachments to the wall of the cavity. Sight is very often damaged or destroyed in the corresponding eye by pressure on the optic nerve, or by its implication in the orbital growth (*see* also Intraocular Tumours, p. 246).

(1.) *Distension of the frontal sinus* by retained mucus causes a well-marked, fixed, usually very chronic swelling, not adherent to the skin, at the upper-inner angle of the orbit. At first it is bony, but when advanced it fluctuates. Its course is usually slow, but acute suppuration may supervene, and the swelling may be then mistaken for a lachrymal abscess (p. 54). There is generally a history of injury. The aim of treatment is to re-establish a permanent opening between the floor of the sinus and the nose. The most prominent part of the swelling is freely opened; a finger is passed up the nostril, and the floor of the distended sinus perforated on the finger by a trochar introduced from above through the incision. A thick seton or small drainage-tube is then passed through the hole so made and brought out at the nostril; it must be worn for several weeks or months.

(2.) *Ivory exostoses* sometimes grow from the walls of the same sinus or from neighbouring parts, beginning comparatively early in life, increasing very slowly and causing absorption of some portions of their containing walls. In thinking of removing these tumours the serious danger of fracturing the cranial walls of their containing cavity, and wounding the dura mater, must be taken into account.

(3.) Tumours encroaching on one or both orbits from the base of the skull, the antrum, the nasal cavity, or the temporal fossa generally admit of correct diagnosis, but their treatment does not belong to the ophthalmic surgeon. The suspicion of tumour on the inner or lower wall of the orbit should always lead the ophthalmic surgeon to an examination of the palate, pharynx, and teeth, of the permeability of each nostril, of the functions of the cranial nerves, of the state of the glands behind the jaw on both sides, and to an enquiry as to epistaxis or discharge from the nose.

(4.) *Pulsating and fluctuating tumours of the orbit and cases of proptosis with pulsation* may be due to:—(*a.*) True aneurism of the ophthalmic artery or one of its branches (but this is very rare). (*b.*) False aneurism from rupture of an artery (or rarely of a true aneurism). Most of these cases are the result of injury to the head or orbit; they vary much in the rapidity of onset of the symptoms and the presence or absence of inflammatory changes in the skin and conjunctiva. (*c.*) In some cases presenting all the symptoms of aneurismal tumour of the orbit, post-mortem examination has disclosed only some cause of compression of the ophthalmic vein at its exit from the cavity, *e. g.* by an aneurism of the internal carotid or phlebitis of the cavernous sinus. (*d.*) Pulsating and erectile tumours well-defined and easily separable have also been met with.

The difficulty of diagnosis in cases of orbital pulsating tumours is thus very great. In examining a case we must note the effect of:—(1) compression of the common carotid on the same side; (2) steady pressure on the eyeball through the closed lids, whether on removing the pressure the former

16

state is *slowly* or *quickly* reproduced; (3) the seat of greatest pulsation, whether the pulsation is strong or weak, the effect of posture, the presence of a bruit heard by the stethoscope or at a distance through the air, and the character of any sound heard by the patient in his own head; (4) pain and inflammatory symptoms and history of injury; pain is often severe in cases of traumatic aneurism with extravasation and orbital inflammation.

Ligature of the common carotid has been practised with good results in a considerable number of cases of pulsating orbital tumour; but the treatment of these cases does not belong to the ophthalmic surgeon.

(5.) A tumour which fluctuates freely, does not pulsate, is free from inflammatory symptoms, and not connected with the frontal sinus, may be a chronic orbital abscess (see also p. 51), a hydatid, or a cyst containing bloody or other fluid and of uncertain origin. An exploratory puncture should be made after sufficiently watching the case, and the further treatment must be conditional. Perfectly clear, thin fluid probably indicates hydatid, and in this case the swelling is likely to return after puncture and the cyst to need removal through a freer opening. The echinococcus hydatid (the only large one infesting man) often contains daughter-cysts, some of which escape puncture. Suppuration may take place around any hydatid.

(6.) Examination leads to the diagnosis of a *solid tumour limited to the orbit.* We have next to determine, if possible, in what part of the cavity the growth began, whether in the eyeball or optic nerve, or in some of the surrounding tissues. We therefore examine the globe for symptoms of intraocular tumour, particularly detachment of retina, cataract and glaucoma (p. 246).

Solid growths independent of the eyeball may arise (*a*) from the *periosteum;* these are firmly attached by a broad base, are generally malignant, and seldom admit of successful removal. (*b*) The *lachrymal gland* is the seat of various morbid growths, including carcinoma (compare p. 51); the diagnosis is often easy, since the chief portion of the growth will be in the position of the gland, and a large part of it can be explored by the finger. Although such a growth is often attached firmly to the orbital wall, its position, lobulated outline, and well-defined boundary will often lead to a correct conclusion. Tumours of the lachrymal gland should always be removed if they are increasing; for we can never feel sure that they are innocent. (*c*) Solid tumours originating in some of the softer orbital tissues, especially the form known as cylindroma, or plexiform sarcoma, occur more rarely. (*d*) Tumours of the optic nerve are rare; they have occasionally been extirpated without removing the globe; they usually cause blindness and neuro-retinitis, but no absolutely pathognomonic symptoms.

When an orbital tumour is found during operation to be adherent to the wall or to infiltrate the tissues around it, chloride of zinc paste should be applied on strips of lint, either at once, or the day after the operation, when oozing has ceased and suppuration has not yet begun. If the periosteum is affected it is to be stripped off and the paste applied to the bare bone. Hæmorrhage from the apex of the orbit, after removal of the globe and tumour, can always be controlled by perchloride of iron and a firm compress.

In every case of suspected primary orbital tumour (unless the growth be quite clearly limited to the lachrymal gland), the question of syphilis must be very carefully gone into. Neither periosteal nor cellular nodes, are common in the orbit, but both varieties occur sometimes, and disappear under proper treatment.

D. INTRAOCULAR TUMOURS

By far the commonest forms are glioma of the retina and sarcoma of the choroid.

Glioma of the retina is always a disease of infancy or early childhood, the patients being generally under two years old when first brought for treatment; it may, however, be present at birth, and may begin as late as the eleventh or twelfth year. It is a very soft small round-celled growth from the granule layers of the retina, and either grows outwards causing detachment of the retina, or inwards into the vitreous; often several, more or less separate, lobules are present. It runs a comparatively quick course filling the eyeball in a few months, spreading by contact to the choroid, and thence to the sclerotic and orbit. It has an especially great tendency to travel back along the optic nerve to the brain; it may cause secondary deposits, in the brain and in the scalp, and more rarely in distant parts. If the eye be removed before either the optic nerve or the orbital tissues are infiltrated, the cure is radical, but in the more numerous cases, where the patient is not seen till what may be called, clinically, the second stage (see below), a fatal return occurs in the orbit or within the skull. Glioma sometimes occurs in both eyes one after the other, and in several children of the same parents.

The earliest symptom is a shining whitish appearance deep in the eye, and the child's mother then soon finds that the eye is blind. As neither pain nor redness are present the case is often not brought for advice until several months after this discovery is made. When the eye begins to protrude, or becomes red and painful, the child is brought. In this (the second) stage there is generally some congestion of the scleral vessels, and a white, pink, or yellowish reflexion from behind the lens (which remains clear), some steaminess of the cornea, dilatation of the pupil and increase of tension; and there may be enlargement or prominence of the eyeball. On focal examination some vessels can generally be seen on the whitish background, and white specks of calcareous degeneration are sometimes present.

Cases are not very uncommon in young children, in which the above appearances are closely simulated by inflammatory changes in the vitreous, often with detachment of the retina, and the differential diagnosis is sometimes impossible. The presence of iritic adhesions, the history of a definite inflammatory attack preceding the white or yellow appearance in the pupil, and the fact that the tension is normal or subnormal are the points in which these cases differ most from glioma; but when there is any doubt the eye should be excised.

Sarcoma of the choroid (including the ciliary body) is a growth of late or middle life, being rarely seen below the age of 35. The majority of these tumours are pigmented (melanotic), some being quite black, others coloured to different degrees in different parts, a few are quite free from pigment. Some are spindle-celled or mixed, others composed of round cells; some are truly alveolar, but in many specimens there is very little connective tissue stroma, and no very defined arrangement of the cells. These tumours are moderately firm but friable; some are very vascular, and hæmorrhages often occur. The tumour generally grows from a broad base, and forms a well-defined rounded prominence, pushing the retina before it; and hæmorrhage or fluid effusion generally takes place round the base of the growth, so that the retinal detachment is usually much more extensive than the tumour. These tumours often grow slowly so long as they are wholly contained within the eye; two, three, or more years may pass before the growth passes out of the eye and invades the orbit. Though this does not usually occur till the globe is filled to distension by the growth, it may happen much earlier, the cells passing out along the sheaths of the perforating blood-vessels, and producing large extra-ocular growths, while the intraocular primary tumour is still quite small. The lymphatic glands do not enlarge, but there is great danger of secondary growths in distant organs and tissues, especially the liver, and this risk is not entirely absent even when the eye tumour is quite small. Hence early removal is of the utmost importance, and a good, though not too confident, prognosis may be given when the optic nerve and tissues of the orbit show no signs of disease.

Symptoms and course.—If the case is seen early, when defect of sight is the only symptom, the tumour can often be seen and recognised by its well-defined rounded outline, some folds of detached retina often being visible near it. There will seldom be either pain or redness, and the pupil, cornea, and eye-tension may be quite natural. But sooner or later the tumour in its growth sets up symptoms of acute or subacute glaucoma with much pain, and causes secondary cataract, and it is now that relief is usually sought (second stage). Unless some part of the tumour happen to be visible outside the sclerotic or project into the anterior chamber, a positive diagnosis cannot now be given on account of the opacity of the lens, although by exclusion we may often arrive at great probability. If the eye be left alone, or iridectomy be performed, glaucomatous attacks and pain will recur and the eye will either enlarge and gradually be disorganised by the increasing growth, which will then quickly fill the orbit and fungate, or a deceptive period of quiet may follow, and perhaps even some shrinking and reduction of tension may occur, after which the growth will make a fresh start and become apparent. It is chiefly in very old patients that this slow course is noticed. Sarcoma is especially likely to form in eyes previously injured, or in eyes already shrunken from previous disease.

Thus it is apparent that in a majority of cases

the diagnosis of choroidal tumour can only be made conjecturally. We suspect tumour and urge excision in the following cases. (1) When an eye that has been blind for some months or years becomes painful, congested, and glaucomatous (there being no glaucoma of the other eye), and particularly if there be secondary cataract. If both be glaucomatous the question of tumour will hardly arise, for the case will then usually be one of primary glaucoma. (2) Similar eyes with normal or diminished tension are best excised as possibly containing tumour. (3) In extensive detachment of retina confined to one eye, without history of injury or evidence of myopia, the patient should be warned, or the eye excised, according to circumstances. The operation of puncturing the sclerotic, used in some cases of detachment, has lately been proposed (Hirschberg) as a means of distinguishing detachment by tumour from separation by fluid only; if a tumour were present the detachment would not subside.

In all suspicious cases the cut end of the optic nerve of the excised eye should be carefully looked at, and if it be pigmented or thickened another piece should be at once removed, and the orbit searched by the finger for evidence of growth; the surface of the eye should also be carefully examined for external growths. When infection of the nerve or orbit is suspected chloride of zinc should be applied as already directed.

In a few rare cases tubercular growths of large size occur in the choroid. The diagnosis is uncertain till after excision, and the treatment differs in no way from that of malignant growths. The patients are generally young.

Tumours of the iris are rare. Melanotic as well as unpigmented sarcomata are occasionally met with, and have been extirpated by removing the affected part through a corneal incision. Sebaceous or epithelial tumours are also seen; they are nearly always the result of transplantation of epithelium, or even of a hair, into the iris through a perforating wound of the cornea.

In rare cases cystic tumours with thin walls are formed in connection with the iris, particularly in eyes operated on for cataract.

Iritis and keratitis punctata may occur in the early stage of a sarcoma of the iris; and there may for a time be difficulty in distinguishing between such a case and syphilitic iritis accompanied by a large nodule (p. 103).

CHAPTER XIX

ERRORS OF REFRACTION AND ACCOMMODATION

When the length of the eye is normal and the accommodation relaxed the percipient or rod and cone layer of the retina is situated exactly at the principal focus of the lens system (*i. e.* of the cornea and crystalline lens); only parallel rays are focussed on the retina, and conversely pencils of rays emerging from this retina are parallel on leaving the eye (Fig. 49). This, the condition of the normal eye in distant

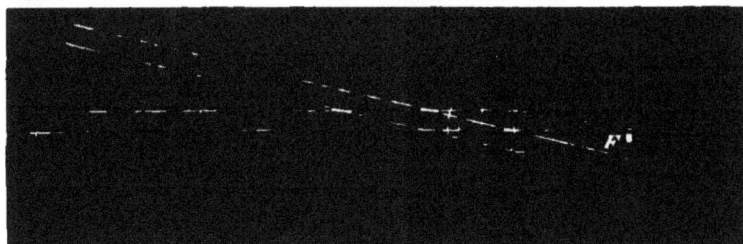

FIG. 49.

vision, is called emmetropia (Em.); and the same eye ceases to be emmetropic when accommodating for a near object. All permanent departures from the condition in which the retina lies at the principal focus are known collectively as ametropia.

In the normal eye the yellow spot of the retina is situated 15 mm. (neglecting fractions) from the posterior pole of the lens, and the posterior focal length of the refractive media (lens system), measured from a certain point in the anterior chamber, is nearly 20 mm. The focal length of the cornea is 31 mm.; that of the lens varies from 43 mm. in Em., to 33 mm. when the eye is accommodated for near objects. The optical centre, or crossing point for the axial rays

of light, practically coincides with the posterior surface of the lens, and is, therefore, about 15 mm. from
the retina.   The length of the visual axis (a line joining the y. s. with the centre of the cornea) is about 23
mm.; the centre of movement of the eyeball is placed
somewhat behind the middle of this axis, and about
6 mm. behind the posterior surface of the lens.

In emmetropia rays from any near object, e. g. divergent rays from *Ob*. Fig. 50, will be focussed behind the

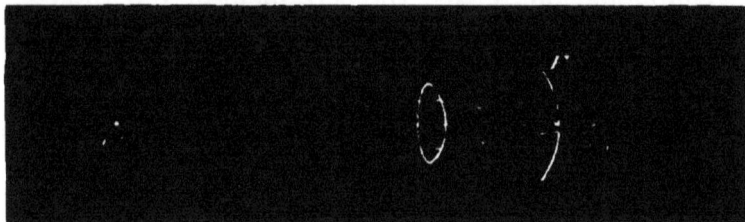

FIG. 50.

retina (at CF), because every conjugate focus of a lens
is further from it than its principal focus.   Reaching
the retina before focussing, such rays will form a
blurred image, and the object *Ob* will therefore be
seen dimly.   But by using accommodation the convexity of the crystalline lens can be increased and its
focal length shortened, so as to make the conjugate
focus of *Ob* coincide exactly with the retina (CF,

FIG. 51.

Fig. 51).   Under this condition the object *Ob* will be
clearly seen, whilst at the same time the image of a
distant object, which in the emmetropic state was
formed on the retina, will now lie in front of it
(F, Fig. 51), and the distant object will appear
indistinct.

In Fig. 50, if the retina, instead of being situated at F, were put back to C F, a clear image would be formed of an object at *Ob*, without any effort of accommodation, whilst objects farther off would be focussed in front of the retina. This state, in which the posterior part of the eyeball is too long, so that, with the accommodation at rest, the retina lies at the conjugate focus of an object at a finite distance, is called Short sight or Myopia (M.).

FIG. 52.

In Fig. 52 the inner line at R is the retina, and F the principal focus of the lens system. Pencils of rays emerging from R will, on leaving the eye, be convergent, and meeting at the conjugate focus R' will form a clear image in the air. Conversely, an object at R' will form a clear image on the retina (R). The image of every object at a greater distance than R' will be formed more or less in front of R, and every such object must, therefore, be seen indistinctly. By exerting accommodation, however, objects nearer than R' will be seen clearly, just as in the normal eye (compare Figs. 50 and 51; see p. 19).

The furthest distance of distinct vision, whether in myopia or other conditions, is called *punctum remotum* (r); the nearest point, *i. e.* with full action of accommodation, is *punctum proximum* (p) (p. 19 (13)).

In Myopia the indistinctness of objects beyond the far point (r) is lessened by partly closing the eyelids. This habit is often noticed in short-sighted people who do not wear glasses, and from it the word myopia is derived.

The distance of r (r', Fig. 52) from the eye will, of course, depend on the distance of its conjugate focus r, *i. e.* upon the amount of elongation of the eye. The greater this elongation, the greater the distance of r from the crystalline lens, the less will be the distance of its conjugate focus r' ($= r$.), the higher will be the myopia, and the greater the indistinctness of distant objects. If the elongation of the eye be very slight, r nearly coinciding with f, r' ($= r$) will be at a much greater distance, and distant objects will be less indistinct. The retinal image formed in a myopic eye is larger than that of the same object formed in a normal eye at the same distance, and so a myopic person can distinguish smaller objects at the same distance than one with normal eyes.

Symptoms of myopia.—In low degrees the patient complains that he cannot see distant objects so well as other people; in moderate and high degrees his complaint is often rather that he can only see distinctly when things are held very close, for objects a few feet off are so indistinct that many such patients neglect them. On inquiry we shall generally find that distant sight was good till from about eight to twelve years of age, that it has been getting shorter since then, and that the defect is either still increasing or has become stationary.

In many cases, no other complaint is made; but in a certain number complications are present. There is often intolerance of bright light, an additional cause for the half-closed lids and frowning expression so often noticed. Aching of the eyes is a very common and troublesome symptom and frequently shows that the myopia is increasing; it is often brought on and always made worse by over use of the eyes, but sometimes is very troublesome when quite at rest, and even in bed at night. One or both internal recti frequently become weakened in myopia, so that convergence of the optic axes for near vision becomes difficult, painful, or impossible, and various

degrees of divergent strabismus result; this occurs
oftenest, but by no means only, in the higher degrees
of M. where r is so near that binocular vision involves
a strong effort of convergence. Double vision is
often complained of when this " muscular asthenopia "
or " insufficiency of the internal recti " is slight or
only occasional, but is seldom present when a constant divergent squint has been established. The
lower degrees of M. are sometimes accompanied by
excessive but involuntary contraction of the ciliary
muscle ("spasm of accommodation ") by which M.
is apparently increased, and the habitual approxi-
mation of objects which thus becomes necessary is
one cause of still further elongation of the eye and
increase of the structural M. Floating specks (*muscæ
volitantes*) are especially common and troublesome in
myopia.

Objective signs and complications. In high degrees
of M. the sclerotic is enlarged in all directions (Fig.
53), the eye often looks too prominent or too large,

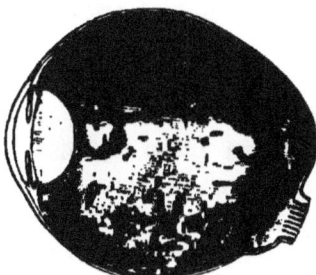

FIG. 53.—Section of a highly myopic eyeball. The retina
has been removed.

and its movements are somewhat impeded. But the
apparent prominence and size of the eye depend on
too many other factors to be safely relied upon as
guides to the diagnosis of myopia (p. 9).

The existence of myopia is made certain by the
ophthalmoscope in three different ways. (1) By
direct examination (p. 36 (2.)), the image of the
fundus formed in the air (Fig. 52) is clearly visible

to the observer, provided he is not nearer to it than his own near point, *p*. It is an inverted image and magnified, the enlargement being greater the further it is formed from the patient's eye, *i. e.* the lower the myopia. For very low degrees this test is not easy to use, because of the great distance (3′ or 4′ *e. g.*), that must intervene between observer and patient. It is easily applied, however, if the image be not more than about 18″ in front of the patient. When with the mirror alone, some of the retinal vessels or a part of the disc are clearly seen at a distance of 2′ or less from the patient, the image is known to be inverted and the eye therefore myopic, if when the observer slightly moves his head from side to side the image seems to move in the opposite direction. The formation of this image may be studied by looking through a convex lens at an object placed a little beyond its principal focus; at a certain distance from the lens a clear inverted image of the object will be seen, which when the observer's head is moved, will move in the opposite direction (compare pp. 37 and 26).

(2.) By indirect examination the disc in M. appears smaller than usual. If, now, the large lens be gradually withdrawn from the patient's eye the disc will seem to increase in size. This test like the former is less applicable in low degrees of M.

(3.) By direct examination no clear view of the fundus is obtained if the distance between patient and observer be less than that at which the inverted aërial image (ᴋ′, Fig. 52) is formed; and as ᴋ′ is necessarily formed at some distance in front of the myopic eye it will be invisible if the observer go close to the patient. Hence, if on going close to the patient the observer cannot, either by relaxing or using his accommodation (see p. 37—9), see any details of the fundus clearly, the patient is myopic (opacities of the media being, of course, excluded). This test is applicable to all degrees of M., but the

complete voluntary relaxation of accommodation,
which is necessary for success, requires some practice.
The tests (1) and (2) are on the whole most generally
useful for beginners.

In a large proportion of cases the elongation of the
eye causes atrophy of the choroid on the side of the
optic disc next to the yellow spot (the apparent
inner side in indirect examination). This change
occurs in the form of a crescentic patch (Fig. 54) of
yellowish-white or greyish colour, whose concavity is

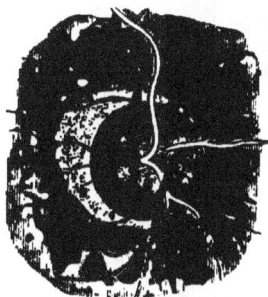

FIG. 54.—Myopic crescent or small posterior staphyloma
(Wecker and Jaeger).

formed by the border of the disc, whilst its convex
side curves towards the yellow spot, and it is commonly known as the "myopic crescent." It is also
called a " posterior staphyloma " because it indicates
a localised bulging of the sclerotic (compare Fig. 53).
It varies in size from the narrowest rim to an area
several times that of the disc, and may form a zone
entirely surrounding the disc (Fig. 55) instead of a
crescent ; there may also be separate spots of atrophy
or diffused thinning of the choroid beyond the bounds
of the crescent, especially in a horizontal direction
towards the yellow spot. As a rule the higher the
myopia the more extensive are these choroidal
changes, but the relation is by no means a constant
one, and occasionally even in high degrees we find no
crescent. Hæmorrhages may occur from the choroid
in the same region, and leave some residual pigment

(pp. 156 and 159).   Owing to the steepness of the
bulging in some cases of M. the disc is tilted and
appears oval because   seen at "three quarter face"
instead of "full face" (Fig. 55).   The disc is some-

Fig. 55.—Large posterior staphyloma (Liebreich).

times markedly pale on the side next the yellow spot
when the staphyloma is large.

There is in myopia a great liability to liquefaction
of, and the formation of hæmorrhagic opacities in,
the vitreous, and, still worse, to detachment of the
retina.   A very large proportion of all the retinal
detachments occur in myopic eyes.   A blow on the eye
often appears to have caused the detachment, though
often not until after a considerable interval.   In high
degrees of M. the lens frequently becomes cataractous,
the cataract generally being cortical and complicated
with disease of the vitreous (p. 146).

Thus we arrive at a sum total of serious difficulties
and risks to which myopic persons are subject, especially when the myopia is of high degree.   It is only
when the degree is low (less than $\frac{1}{20}$th), and the
condition stationary that the popular idea of " short
sight " being " strong sight " is at all borne out, or
that the later onset of presbyopia counter-balances the
disadvantages of bad distant vision.

Causes.—The tendency to M. is often hereditary, several near relatives often being affected.
But in most cases it is acquired by the prolonged
use of the eyes in looking at objects held at a very
short distance ; the strain on the internal recti coun-
terbalanced, it may be, by a corresponding tension on
the external recti, is believed to act by slightly
bulging out the unprotected posterior pole of the
sclerotic. The concomitant tension of the accommodation probably aids by bringing on congestion of
the uveal tract (as it certainly appears to do of the
disc), and thus predisposes to softening and yielding
of the tunics ; to this congestion the habit of stooping
over the book or work also contributes by retarding
the return of blood.   It is evident that if the disease
be once started by such causes they will tend powerfully to increase it.   Myopia is a disease of childhood
and early adult life, seldom increasing after about
twenty-five, unless under special circumstances ; but
general enfeeblement of health, as after severe illness
or prolonged suckling, seriously increases the risk of
its increase.   Any condition of imperfect sight in
childhood in which better sight is gained by holding
all objects very close is likely to bring on M.; and
so we find it disproportionately common amongst
those who from childhood have suffered from
corneal nebulæ, partial (especially lamellar) cataract,
heredito-syphilitic choroiditis, or a high degree of
astigmatism (p. 273).

The *treatment* is divisible into (1) prophylactic
and (2) remedial. 1. Much may be done to prevent M., or to check its increase when it has begun,
by regulating the light, the books, and the desks
used by children, so as to remove the temptations to stooping.   Children should not be allowed
to read or work by flickering or dull light ; and as
we write and read from L. to R., it is best, whenever
possible, to sit so that the light comes from the left,
and throws the shadow of the pen towards the right

and away from the object looked at. A myopic child should be prevented from fully indulging his bent, which is generally strong, for excessive reading.

2. Suitable glasses are to be worn (a) to give clear vision of distant objects, i.e. for making the eye emmetropic; (b), to allow of reading and working at a greater distance. The former is important only for educational purposes, that the patient may see what is about him as clearly as other people; as the strain on the internal recti in near vision ceases when the gaze is directed in the distance, whether vision be distinct or not, glasses for distant vision have no effect on the progress of the myopia, and we do not insist upon them on medical grounds. But by somewhat increasing the distance of the natural far point (r), from the eyes we lessen the tension on the internal recti in near vision, and diminish the temptations to stooping and to reading by bad light, and so help to check the progress of the disease; hence glasses for near work are important in the higher degrees of myopia (from $\frac{1}{12}$ upwards) in early life. When the disease has been stationary for years, however, we may generally leave even this point to the patient's own choice.

For either purpose we must first measure accurately the degree of M. (a) In Fig. 56 let r be the far point, and let it be 10″ in front of the patient's eye, so that he can see nothing clearly at a greater distance than 10″. He is required to see distant objects (objects seen under approximately parallel rays) clearly. A concave lens is interposed of strength sufficient to give to parallel rays a degree of divergence, as if they came from r. The focal length of this lens will be the same as its distance from r; and, as it is placed close to the eye, its focal length will be very nearly the same as (a little shorter than) the patient's far point. Therefore, if we measure the distance of r from the patient's eye a lens of

nearly the same (usually rather shorter) focal length*
will fully neutralise his myopia.   In practice he will

Fig. 56.

choose a rather higher lens if the M. be uncomplicated;
whilst, if owing to complications (p. 256) there be
considerable defect of vision, M. will often prove
to be less than, judged by this test, we should expect.
Hence the rule is to begin the trial with a lens considerably weaker than the one which, judging by the
greatest distance of distinct vision, we expect the
patient to choose, and to try successively stronger
ones till the best result is reached.   The weakest
concave glass which gives the best attainable sight
for the distant test types (p. 18) is the measure of
the M., and this glass, *but not a stronger one*, may be
safely worn for distant vision.

(*b*) Often a glass is needed with which the patient
will be able to read or sew at a greater distance than
his natural far point. Theoretically the fully correcting glass (*a*) would suit, since it gives to all rays a
course which, in relation to the retina, is the same as
that of the rays entering a normal eye.   But this glass
cannot safely be given in the higher degrees of M.

* It is sometimes stated that the glass chosen for distance is
rather *weaker* than is indicated by the distance of *r* from the
crystalline lens, the accommodation causing an apparent increase
of M.   This is true only in low degrees of M., and not always
even in them ; a large number of the patients choose a somewhat
stronger lens than is indicated by *r*, a glass stronger, by the
distance between its own central point and the optical centre
of the eye.

The lens which fully corrects the myopia diminishes the size of the retinal images so much that the patient is tempted to increase them again by approaching the object nearer; again, the accommodation is defective, at least in the higher degrees of M., and, as the fully correcting lens requires full accommodation, it will lead to over-straining if the function be weakened, and so cause discomfort, if nothing worse. For these two reasons the rule is to give, for near work, a glass which will diminish the myopia, but not fully correct it.

Let M. be $\frac{1}{5}$, then r will be at (nearly) 5″ from the eye. A glass is required with which the patient shall be able to read at 12″, or which shall remove r from 5″ to 12″, *i.e.* shall leave the patient with M. $\frac{1}{12}$. We must, therefore, correct the difference between $\frac{1}{5}$ and $\frac{1}{12}$ ($\frac{1}{5}-\frac{1}{12}=\frac{1}{8\frac{1}{3}}$); and a concave lens of $8\frac{1}{3}$″ focal length will make rays from 12″ diverge as if they came from 5″. But even this partially correcting lens may diminish the images so much that, if vision is imperfect from extensive choroidal changes, reading at the prescribed distance will be impossible or so difficult to learn that the patient will give it up and bring the object nearer again; it is, therefore, often advisable, even for partial correction, to order a weaker lens than is optically correct.

Aching from preponderance of the external over the internal recti (insufficiency of the internal recti, pp. 188 and 253), if not cured by partially correcting glasses, is often best treated by division of the external rectus of one or both eyes. This operation may always be done when there is a marked divergent squint even if the squint be variable.

Prismatic spectacles (p. 22), the bases of the prisms being towards the nose, are found by some practitioners to be very serviceable for reading, in cases of muscular insufficiency. By deflecting the entering light towards their bases the prisms give to rays from a certain near point a direction as if they came from a greater distance, and thus lessen the need for convergence of the optic axes. The prisms may be combined with concave lenses.

Myopia may be caused not only by elongation of the eye, but by increase of the refractive power of the cornea or lens. Thus, in conical cornea (p. 86) the curvature of the central part of the cornea is increased (*i. e.* its focal length shortened), and though there be no posterior elongation of the eye the principal focus lies in front of the retina, often very far in front, and the eye is consequently myopic. In commencing cataract, the lens sometimes becomes either more convex by swelling, or more highly refractive, and the result is a certain degree of myopia. M. is sometimes simulated in H., and actual M. increased by needless and uncontrollable action of the ciliary muscle (see p. 253).

HYPERMETROPIA

Hypermetropia (H.) is optically the reverse of myopia. It is one of the commonest conditions we have to treat. The eyeball is too short, so that when the accommodation is relaxed the retina lies within the principal focus of the lens-system. As rays from an object within the principal focus of a convex lens emerge from the lens divergent, so pencils of rays leaving a hypermetropic eye are divergent; and, conversely, only rays already convergent can be focussed on the retina. H. always dates from birth and does not afterwards increase, except slightly in old age (p. 263). But it may diminish and if neglected even give place to M. by the elongation of the eye. In Fig. 57 the curved line representing the retina is in front of F. (compare Fig. 49). Consequently, parallel rays will, after passing through the lens, meet the retina before focussing and form a blurred image, whilst divergent rays, meeting the retina still further from their focus, will form an even worse image (Fig.

50) ; and hence, neither distant nor near objects will be seen clearly. But by using accommodation the focal length can be shortened until the focus lies

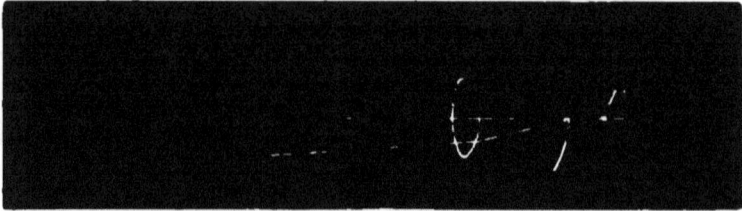

FIG. 57.

exactly upon the retina (Fig. 58), and distant objects are clearly seen ; and additional accommodation will give also distinct vision of near objects (compare

FIG. 58.

Fig. 51). A little consideration will show that the competence of the ciliary muscle to give these results will depend in any given case, 1st, on the degree of advancement of the retina in front of F.,

FIG. 59.

i.e. on the degree of shortening of the eye; and 2nd,

on the strength of A., *i. e.* on the extent to which the focal length of the lens can be shortened.

Fig. 59 may be taken for a section of a highly hypermetropic eye, the rays emerging from which are divergent. The shortening here shown (about 5 mm.) would give a degree of H., which for its correction would need accommodation equal to more than a 2″ lens; such a high degree is never seen, except when the crystalline lens is absent, as after operations for cataract (p. 150). The image formed on the retina of a hypermetropic eye is smaller than the retinal image of the same object placed at the same distance from a normal eye (compare p. 252).

In old age the crystalline lens becomes flatter and less refractive, and, therefore, an eye originally emmetropic is now unable to focus parallel rays, on the retina, and becomes slightly hypermetropic; this is termed acquired hypermetropia.

Symptoms and results of hypermetropia.—The *direct* symptoms are due to the insufficiency of the accommodation; distinct vision of all objects whatsoever, not only near, but distant, requires accommodation proportionate to the degree of shortening of the eye. In a given case, A. being relaxed, let the rays on leaving the eye diverge, as if they proceeded from a point, 10″ behind it (strictly 10″ behind the crystalline lens). If parallel rays be passed through a convex lens of 10″ focal length before entering the eye, they will receive such a degree of convergence as would focus them 10″ behind it, and the additional refraction of the lens-system of the eye will focus them upon the retina; this end can be equally gained by using A., so as to shorten the focal length of the crystalline lens to a corresponding extent.

If H. is slight or moderate and A. vigorous, no inconvenience will be felt either for near or distant vision. But if A. have been weakened by disease or ill health, or have failed with age, the patient will complain that he can no longer see near objects clearly for long together; that the eyes ache or water, or that everything " swims " or becomes " dim," after reading or sewing for a short time (accommodative asthenopia). In children, the complaint will sometimes be of watering and blinking rather than of dimness, because they are not old enough to describe their symptoms. There is not usually much complaint of defect for distant objects. Many slight or moderately H. patients find no inconvenience till 25 or 30 years of age when accommodation has naturally declined by nearly one half (p. 279). Women are often first troubled after a long lactation, and men when they have had to work hard for examinations or in the office, or are suffering from chronic exhausting diseases.

But in the higher degrees of H. the greater part of the accommodation is always needed from childhood upwards for distant sight; and even the strongest effort does not suffice to give good vision of near objects, which consequently such a person never sees clearly. Such patients often partly compensate for the incurable dimness of near objects by bringing them still nearer, and so increasing the size of the retinal images. This symptom may be mistaken for myopia, but can be distinguished by the want of certainty in the patient's manner of placing his book, and by the fact that he often cannot at any distance whatever see the print easily nor read fluently. In these high degrees of H. even distinct distant vision is not constantly maintained, the patient often being content to let his accommodation go and see surrounding things dimly, except when his attention is roused.

In considerable H., as age advances, a point is reached at which the accommodation no longer suffices even for distant, and much less for near, vision. Such patients will tell us that they took to glasses for near work comparatively early, but will add that lately the glasses have not suited, and that they are unable to see clearly either at long or short distances. Ophthalmoscopic examination shows no change excepting H., and suitable convex glasses at once raise distant vision to the normal. Occasionally photophobia, slight conjunctival irritation and redness are present in H., but the former symptom is less common than in myopia (see also p. 188).

The most important *indirect result* of H. is convergent strabismus. To understand how this arises we must remember that there is a certain constant relation between the action of the ciliary muscles and of the internal recti, that the accommodation can be exerted only to a very limited degree without convergence of the optic axes, and that for every degree of accommodation there is in the normal state a constant amount of convergence (p. 19). In H. accurate near sight needs, as we have seen, an excess of accommodation, thus *e.g.* with H. of $\frac{1}{16}$, clear vision of an object at 16″ will require as much A. as vision at 8″ by a normal eye, and this A. cannot be exerted without converging for 8″ (or nearly so). Such a person, therefore, has to do two things at once—to look at an object distant 16″, and to make his optic axes meet at 8″. The former object is gained by accommodating for the object 16″ off, the second by converging the visual axis of one eye so as to meet that of the other at a distance of 8″; the converging eye will, therefore, not look at the object of regard, but will squint inwards, its axis cutting that of the first eye at 8″ instead of at 16″.

This convergent strabismus generally comes on early in childhood, as soon as the child begins to look attentively and use its accommodation vigorously in regarding near things. It is called *concomitant squint*, since it arises in companionship with the accommodation. In examining cases we shall be struck by finding that; (1) in some the squint is noticed only when A. is in full use, that it appears and disappears under observation according as the child fixes its gaze on a near object or looks into space ; (2) in others the squint is constant, but is more marked during strong A.; (3) it is constant, invariable, and of high degree. The first variety is known as *periodic squint*. In most cases the squint always affects the same eye, but some patients squint with either eye indifferently (*alternating squint*). The constancy of the squint to one eye which is noticed in most cases, is generally accounted for by some original defect of the eye itself (such as a higher degree of H., or As., or a corneal opacity), which leads to its fellow being chosen for distinct sight. The squint causes diplopia, which is homonymous (p. 285), and to avoid this inconvenience, patients for the most part soon learn to ignore (or " suppress ") the image formed in the squinting eye, the result usually being that this eye becomes defective and often almost blind (p. 230). This power of suppressing the false image is learnt most easily in very early life. In alternating squint no permanent suppression occurs, and consequently both eyes remain good.

It will soon be noticed that squint is not present in every case of H. In very low degrees the necessary extra accommodation can be used without any extra convergence (relative accommodation, p. 19). In very high degrees, on the other hand, the effort needed for distinct vision, even of distant things, and *à fortiori* of near objects, is so great, that the child often sacrifices distinctness to comfort and binocular vision, using only so much accommodation as can be employed without over-convergence, and hence he does not squint. In a few cases a concomitant squint spontaneously disappears as the child grows up, a fact, perhaps, explained by an increased power of dissociating A. from convergence, or, perhaps, by a diminution of H. or even actual onset of M. from elongation of the eye.

The *treatment* of H. consists in removing the necessity for overuse of A. by prescribing convex spectacles which, in proportion to their strength, supply the place of the increased convexity of the crystalline lens induced by A. In theory, the whole H. ought to be corrected by glasses in every case, and the eye be rendered emmetropic when its accommodation is at rest. But in practice we find it often better to give a weaker glass, at least, for a time.

If the accommodation be in abeyance (paralysed by atropia) parallel rays being no longer focussed on the retina, but behind it, distant objects (and *à fortiori* near objects) will be indistinct. If, now, the patient look at distant objects through a convex lens of strength just sufficient to shorten the focus until it falls on the retina, distant objects will be seen clearly (Fig. 58). The focal length of this lens is the measure of the H. ; the patient has H. $\frac{1}{10}$th if a convex lens of 10″ focal length is necessary for this purpose.

But if the accommodation be intact, the patient having constantly to use it for distant sight, is often unable to relax it fully when a corresponding convex lens is placed in front of the eye ; he will relax only a part, and this part will be measured by the strongest convex lens with which he can see the distant types clearly. The part of the H. which can be detected by this test is called "manifest" (H. m.). The part remaining undetected, because corrected by the involuntary use of accommodation, is latent (H. l.). The sum of the H. m. and H. l. is the total, (H.).

Now, most people can habitually use some A. for distance (and a corresponding excess for near vision) without inconvenience, and hence the full correction of H. is by no means always needful nor even agreeable to the patient. In many cases the correction of the H. m. is enough to relieve the asthenopic symptoms, at any rate, for a considerable time ; but we often find that after wearing these glasses for some weeks or months the symptoms return, and a fresh trial will then show a larger amount of H. m., which must then again be corrected by a corresponding increase in the strength of the glasses. This process may have to be repeated several times until after a few months the total H. becomes manifest, and may be corrected. This method is most suitable for adults to whom the use of atropine for paralysing A. and allowing the immediate estimation of the total H. is inconvenient or impossible; or for whom the glasses which correct the total H., as estimated by the ophthalmoscope (p. 37), without atropisation, are found, if ordered at once, to be inconveniently strong. But for children there is seldom any gain and often no little inconvenience from following this gradual plan; with them the better way is to estimate the total H. either by the ophthalmoscope, or by glasses after the full use of atropine, and to correct nearly the whole of it.

To examine for H.—(1.) For H. m. Note the patient's vision for distant types at 20′, then hold in front of his eye a very weak convex lens ($+ \frac{1}{80}$ or $+ \frac{1}{40}$), and if he sees as well or better with it go to the next stronger lens, and so on until the strongest has been found which allows the best attainable distant vision; this lens is the measure of the H. m.

(2.) For H. (total).—The easiest and most certain plan is to direct the patient to use strong atropine drops (sulphate of atropine gr. iv to ℥j) three times a day for at least two days, and then to test his distant vision with convex glasses. As in 1, the strongest lens which gives the best attainable sight is the measure of H.

If the observer be skilled enough he may, as stated at p. 37, estimate H. with almost as great accuracy by a refraction ophthalmoscope as by trial lenses, and this plan is often almost indispensable with children who are too young to give good answers, or are backward on account of their defective eyes. The total, or nearly the total, may often be found in this way without atropine if the examination be made in a very dark room, for then A. is generally quite relaxed, however persistently it may have acted when the patient was able to look attentively at objects in the light. But it is often better in practice to use atropine before making this estimation.

The next question is, whether the glasses are to be worn always, or only when the accommodation is specially strained (in near work). They are to be worn constantly (1) whenever we are attempting to cure a squint by their means; (2) in all cases of high H. in children, whether with or without strabismus. They may be worn only for near work by most of the patients who come under care for the first time as young adults, in whom the H. is, as a rule, of moderate or low degree. Elderly persons require two pairs—one neutralising the H. m. for distant vision—the other a stronger pair, neutralising the presbyopia also (see p. 278).

Treatment of hypermetropic (concomitant) squint

(1.) If the squint be only present during accommodation for near objects (periodic squint), it can be cured by the constant use of spectacles which correct the total H.

(2.) The same may be done in some cases where the squint, though constant, varies in degree, being greater during accommodation for near than for distant objects. It is best to atropise fully all such cases, and if then the squint disappears, or is much lessened, glasses will cure it. We shall, however, often be disappointed to find the squint as marked as ever, even with complete paralysis of accommodation, and operation is then, as a rule, advisable without delay.

(3.) If the squint be constant in amount and

of long standing, operation is always necessary. As the squinting eye is often very defective (p. 266), binocular vision is by no means always obtained after operation, and the removal of the deformity is then often all that can be expected. Only one internal rectus should be divided at a sitting. At the end of a fortnight, if the squint still be considerable, but not otherwise, the operation is performed on the other eye.

ASTIGMATISM

In the preceding cases (M. and H.), each of the refracting surfaces of the eye (the front of the cornea and the two surfaces of the lens) has been regarded as a segment of a sphere. All the rays of a cone of light which issue from a circular spot and pass through such a system are (neglecting "spherical aberration") equally refracted, and meet one another at a single point—the *focus* of the lens.

For if such a cone of incident light be looked upon as composed of a number of different planes of rays situated radially around the axis of the cone, the rays situated in any plane (say the vertical) will, after passing through the lens-system, meet behind it at its focus; those forming any other plane (as the horizontal) will meet at the same point, and the same will be true of all the intermediate planes.

But if the curvature, and, therefore, the refractive power, of any of the surfaces (as, for instance, of the cornea) be greater in one meridian, say the vertical, than in the horizontal, the vertical-plane rays will meet at their focus, whilst the horizontal-plane rays at the same distance will not yet have met, and if received on a screen will form a horizontal line of light. If the intermediate meridians had regularly intermediate focal lengths they would form, at the same place, lines of intermediate lengths, and the image of the spot of light, if caught on a screen at this distance, would form a horizontal oval. To a retina receiving such an image the round point of light would appear drawn out horizontally. Such an eye is called astigmatic, because unable to see a point as such, all points appearing drawn out more or less into lines.

A little reflection will show that in the same case, at the focal point of the horizontal-plane rays, the rays of the vertical plane will already have met and crossed, and that the image at this point will form a vertical oval.

If the screen be placed midway between these two extreme points the image will be circular but blurred, because the vertical plane rays will have crossed, and begun to separate, while the horizontal ones will not yet have met, and both will be equally distant from the axis. The distance between the foci of two planes at right angles to one another (or the difference between their focal lengths) is called the *focal interval*, and represents the degree of astigmatism.

The astigmatism of the eye may be *regular* or *irregular*. In *regular astigmatism* the meridians of greatest and least refractive power, which are called the principal meridians, are always at right angles to each other; the intermediate meridians are of regularly intermediate focal lengths; and every meridian is nearly a segment of a circle. Of the principal meridians the most refractive (the one with shortest focal length) is, as a rule, vertical or nearly so, and the least refractive, therefore, horizontal or nearly so. The cornea is the principal seat of this asymmetry, but the crystalline lens is also astigmatic, though to a much less degree, and its meridians of greatest and least curvature are usually so arranged as in some degree to neutralise those of the cornea, so that it partially corrects the corneal error.

Regular astigmatism is remedied by supplying a lens which equalises the refraction in the two principal meridians. Such a lens must be a segment of a cylinder instead of, like an ordinary lens, a segment of a sphere. Rays traversing a cylindrical lens in the plane of the axis of the cylinder are not refracted, since the surfaces of the lens in this direction are parallel ; but rays traversing it in all other planes are refracted more or less, and most in the plane or meridian at a right angle with the axis.

Irregular astigmatism may be caused either by irregularities of the cornea, arising from ulceration or cornical cornea (p. 86), or by various conditions of the crystalline lens, such as differences of refraction in its various sectors, tilting or lateral dislocation of the entire lens, so that its axis no longer corresponds, as it should do, with the centre of the cornea. Irregular astigmatism is seldom much benefited by glasses.

Returning to regular astigmatism, it will be seen that the optical condition of the eye will depend upon the position of the retina in respect to the focal interval. In the following diagram (Fig. 60) let the most refracting meridian be vertical and its focus be called *a*, the least refracting meridian horizontal and its focus, *b*. The astigmatism is necessarily represented as caused by altered *position of the retina* in different planes, instead of by altered *curvature of the cornea* in different planes ; the diagram is, of course, only intended to aid the comprehension of the prin-

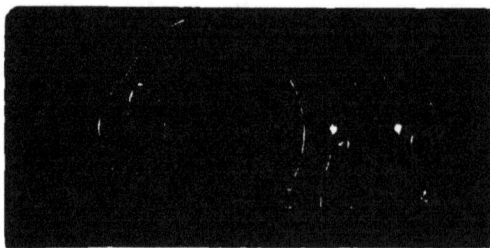

FIG. 60.

ciple. (1.) Let *a* fall on the retina (1, Fig. 60), and *b*, therefore, behind it. There is Em. in the vertical meridian, and therefore H. in the horizontal meridian. This is simple hypermetropic astigmatism. (2.) Let b fall on the retina (2, Fig. 60), and a in front of it. The horizontal meridian is, therefore, Em., and the vertical meridian M. Simple myopic astigmatism. (3), Let a and b both lie behind the retina (3, Fig. 60). There is H in both meridians, but more in the horizontal than the vertical meridian. Compound hypermetropic astigmatism. (4.) a and b are both in front of the retina (4, Fig. 60). There is M. in both meridians, but more in the vertical than the horizontal. Compound myopic astigmatism. (5.) a is in front of the retina, and b behind it (5, Fig. 60). There is M. in the vertical and H. in the horizontal meridian. Mixed astigmatism.

The general symptoms of astigmatism are of the same order as those caused by the simpler defects of refraction, but attention to the patient's complaints or observation of the manner in which he uses his eyes will in the higher degrees often give the clue to its presence. Low degrees, especially of simple hypermetropic astigmatism, often give rise to no inconvenience till rather late in life. As. is most commonly met with in connection with H., because H. is so much commoner than M. But it is said to occur with greater *relative* frequency in M., when for various reasons it may, if not of high degree, be readily overlooked unless specially sought for. The higher grades of As. cause much difficulty to the patient, since no objects can be seen clearly; and ordinary glasses, though of some use if the As. be compound, are useless if it be simple. As. is always to be suspected if, with the best attainable spherical glasses, distant vision is less improved than it ought to be (supposing, of course, that no other changes are present to account for the defect). No definite rule can be laid down as to the degree of defect which should raise the suspicion of As.; indeed, in the higher degrees

18

of even simple M. and H., acuteness of vision is often below normal (pp. 256, and 231 (2.)).

Astigmatism may be measured either by trial with glasses, or by ophthalmoscopic estimation (p. 89) of the refraction of the retinal vessels in the two chief meridians. The latter is the more difficult. A comparatively easy qualitative test is found in the apparent shape of the disc, which, instead of being round, is more or less oval. In the erect image the long axis of the oval corresponds to the meridian of greatest refraction, and is therefore as a rule nearly vertical (Fig. 61, and p. 271).

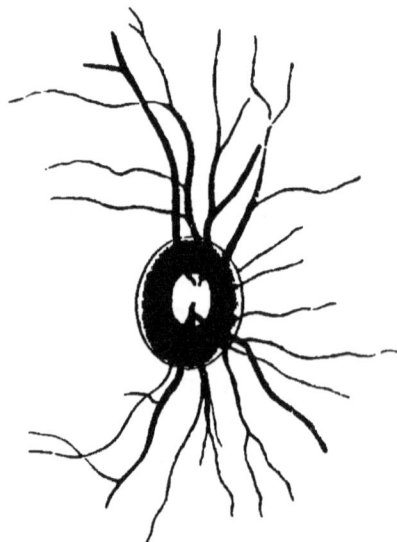

FIG. 61.—Erect image of disc in Astigmatism with meridian of greatest refraction nearly vertical (Wecker and Jaeger.)

In the inverted image (Fig. 62) the direction of the oval is at right angles to the above, provided that the ocular lens (object glass) be nearer than its own focal length to the eye. Astigmatism is suspected when, in the erect image, an undulating retinal vessel appears clear in some parts, and indistinct in others, an appearance which may be taken for retinitis if the examination be confined to the erect image. It may be imitated by looking at a wavy line through a cylindrical lens.

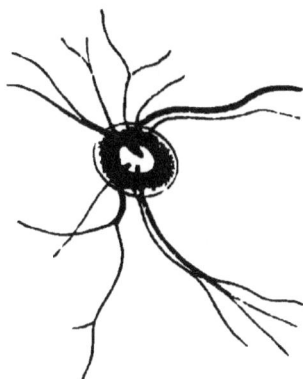

FIG. 62.—The same disc, seen by the indirect method (Wecker and Jaeger).

In the indirect examination the shape of the disc changes on withdrawing the lens from the patient's eye. It will be remembered that in M. the image increases as the lens is withdrawn (p. 254), and that in Em. its size remains the same. In a case of simple myopic astigmatism in the vertical meridian, therefore, that dimension of the disc which is seen through the vertical meridian will enlarge on distancing the lens; from having been oval horizontally, when the lens was close to the eye, it becomes first round and then oval vertically on withdrawing the lens. In the other forms of astigmatism the same holds true; the image enlarges, either absolutely or relatively, in the direction of the most refracting meridian.

The subjective tests for As. are very numerous, but all depend on the fact, that if an astigmatic eye looks at a number of lines drawn in different directions, some will be seen more clearly than others. The form of this test is not a matter of great consequence, provided that the lines are clear, not too fine, and easily visible with about half the normal V. at from 10′ to 20′. The forms resembling a clock face with bold Roman figures at the end of the radii are most convenient, and I prefer the pattern* recommended by Mr Brudenell Carter to any other that I have used. On this face is a broad white "hand," painted with three parallel black lines, which can be turned round into the positions of best and worst vision.

The easiest case for estimation is one of simple H. As., in which the eye is atropised. Many cases of simple M. As. are quite as easy to test. In a given case let the eye be Em. in the vertical meridian, and H. in the horizontal. With the accommodation paralysed, rays refracted by the vertical meridian will be accurately focussed on the retina, whilst the focus of those refracted by the horizontal meridian will be behind the retina (Fig. 60, 1), and will consequently form on the retina a blurred image. Now the rays which strike in the plane of the vertical meridian are those which come from the borders of horizontal lines; hence the patient under consideration will see the lines at a distance of 20′ quite clearly when the hand is horizontal, except their ends, which will be blurred. The rays which strike in the plane of the horizontal meridian are those which proceed from the sides of vertical lines, and as this meridian is hypermetropic the lines in the hand, when placed vertically, will be seen indistinctly, except their ends, which will be sharply defined. We now leave the "hand" vertical, and test the refraction for the lines in this position, i. e. for the horizontal meridian, in the ordinary way (p. 268), and find, e. g. that with + 16 they are seen most clearly, though not perfectly. On substituting for the spherical glass + 16 cylinder with its curvature horizontal (i. e. its axis vertical) the lines of the hand and all the figures on the clock will be seen perfectly : the vertical lines and figures being seen through the horizontal meridian corrected by the cylinder lens; the horizontal figures through the ver-

* Made by Carpenter and Westley.

tical meridian, since the rays which pass through the cylinder in this direction are not refracted.

In a case of simple M. As. in the vertical meridian the lines of the hand will be dull or invisible when horizontal, whilst when vertical they will be clear. On trial a concave cylinder will be found, which, with its curvature vertical (axis horizontal), makes the lines of the hand quite clear when horizontal, and all the figures quite plain.

The cases of compound and mixed astigmatism are less easily detected and dealt with. It is generally best to test them first with spherical glasses for the distant types, and having found the glass which gives the best result, to test the astigmatism in the same way as in the simple cases just described, but with the addition of the chosen spherical lens.

We may use, instead of a cylindrical glass, a narrow slit in a round plate of metal, which can be placed in the direction of either of the chief meridians, and we then find the spherical glass with which in each meridian the patient sees best. One chief meridian may be ascertained by finding the position of the slit in which sight is best with the spherical glass chosen in the preliminary examination, and the other meridian by finding the glass which gives the best result with the slit at a right angle to the former direction.

Another method (that of Javal) consists in making the patient highly myopic for the time being, by means of a convex lens (unless he be myopic already), then accurately finding his far point for the least myopic meridian, and, lastly, finding the concave cylinder which is needed to reduce the opposite meridian to the same refraction. A special apparatus is needed.

Whatever means be employed, the degree of astigmatism is expressed by the difference between the glasses chosen for the two opposite lines, or by the cylinder chosen as giving the best result for the lines or for the distant types; and when cylindrical glasses are ordered the whole of the astigmatism should be corrected.

Vision is, however, often defective in astigmatism, and in the high degrees we are often obliged to be content with a very moderate improvement at the time of examination. This is to a great extent due to the retina never having received clear images, in fact to want of practice in accurate perception (p. 231); and the sight often improves markedly after proper glasses have been worn for some months. Very much also depends, in the trial, on the general intelligence of the patient; some persons are far more apprecia- tive of slight changes in the power of the glass or in the direction of the axis of the cylinder than others, and this quite apart from the absolute acute- ness of sight.

It is not usually necessary to correct astigmatism to less than $\frac{1}{40}$; but exceptions to this rule are not uncommon, some patients deriving marked relief and benefit from the correction of much lower grades.

Unequal refraction in the two eyes (an-iso-metropia). It is extremely common to find a difference between the two eyes, one being more hypermetropic, more myopic, or more astigmatic than its fellow, or one being normal, while the other is ametropic. When the difference is small and acuteness of vision good in both (see p. 231), the refraction may with advantage be equalised by giving a different glass to each eye; but equalisation is seldom possible if the difference is more than will be neutralised by a glass of 36″ focal length. In higher degrees, however, especially of myopia, advantage is sometimes gained by partial equalisation. When no attempt is made to har- monise the eyes the spectacles ordered should be those which suit the *less* ametropic eye.

PRESBYOPIA

Presbyopia (Pr.) (old sight, often called "long sight") is the result of the gradual recession of p which takes place as life advances, and which causes curtailment of the range or amplitude of A. From the age of ten (or earlier) onwards, p is constantly receding from the eye. When it has reached 8″ (when clear vision is no longer possible at a shorter distance than 8″) presbyopia is said to have begun. The standard is arbitrary, 8″ having been fixed by general agreement as the point beyond which p cannot be removed without some inconvenience, the point where age begins to tell on the practical efficiency of the eyes unless glasses are worn.

In the normal eye this point is reached soon after forty, and the rate of diminution is so uniform in most cases that the glasses required may often, if necessary, be determined merely from the patient's age.

Thus, at 45, the glass required to bring p. to 8″ is about $\frac{1}{36}$ (1 D.)

,,	50	,,	,,	$\frac{1}{18}$ (2 D.)
,,	55	,,	,,	$\frac{1}{12}$ (3 D.)
,,	60	,,	,,	$\frac{1}{9}$ (4 D.)
,,	65	,,	,,	$\frac{1}{8}$ (4·5 D)
,,	70	,,	,,	$\frac{1}{6\frac{1}{2}}$ (5·5 D.)
,,	75	,,	,,	$\frac{1}{6}$ (6 D.)
,,	80	,,	,,	$\frac{1}{5}$ (7 D.)

But as allowance has to be made for any error of refraction (H. or M.), and as there are exceptions to the rule even for normal eyes, it is unsafe in practice to rely on age as anything more than a general guide.

The slow failure of accommodation causing Pr. depends upon senile changes in the lens, by which it becomes firmer and less elastic, and therefore less responsive to the action of the ciliary muscle. There can be little doubt, however, that failure of the ciliary muscle itself, or of its motor nerves, also forms an important factor in some cases, particularly when Pr.

comes on comparatively early and very quickly (pp. 199 and 206).

As Pr. depends on a natural recession of the near point it occurs in all eyes whether their refraction be Em., M., or H. In M., however, Pr. sets in later than in a normal eye, because for the same amount of accommodation p is always nearer than in the normal eye. In H., on the contrary, Pr. is reached sooner than is normal because H. has to be neutralised before any A. is available for near vision, and therefore, for the same amount of A., p is always further than in the normal eye. The only cases in which Pr. cannot occur are in M. of more than $\frac{1}{8}$. If a patient has M. $\frac{1}{7}$, his far point r is at $7''$, and though, with advancing years, p will recede to $7''$, it cannot go further, and the patient therefore never becomes presbyopic; the only change for him will be the loss of power to see clearly at less than $7''$; he will be able to see clearly at $7''$, but neither nearer nor further.

Treatment.—Convex spectacles are given, with which the patient can read at $8''$. The lens required can be approximately found by measuring the distance of p when unaided by glasses, and finding the difference between this and the required p ($8''$); or, to put it in another way, by measuring the existing A., and taking the difference between it and the required A. Thus, in a given case p unaided lies at $16''$; the patient has $\frac{1}{16}$ of A., or A. is equal to a convex lens of $16''$ focal length. It is required to be $\frac{1}{8}$, or equal to a lens of $8''$ focal length. The lens needed will be $\frac{1}{8} - \frac{1}{16} = \frac{1}{16}$; a lens of $16''$ focal length will bring p from $16''$ to $8''$.

In practice it is always proper to examine for H. or M., by taking the distant vision and trying the patient for Hm. (p. 268) and M. (p. 258). If Hm. be found, arm the patient with the glasses which neutralise it and make him Em., and then add to them the glasses required to bring p to $8''$.

In prescribing for Pr. we must often be content

with rather less than full correction. For instance,
if A. be almost entirely lost, p is practically removed
to r, and the glass which will bring p to $8''$ will also
bring r to the same, or nearly the same point, and
the patient will be able to see clearly only just at
$8''$. Now $8''$ is too near for convenient sustained
vision, and such patients will prefer a glass which
gives them $p = 10''$, $12''$, or $15''$, though in choosing
it they sacrifice some degree of sharpness of sight.
The difficulty experienced by these patients in reading with glasses which give $p=8''$ or $10''$, depends on
the unaccustomed strain which is thereby thrown on
the internal recti ; and it may be removed or lessened
by adding to the convex glasses prisms, the bases of
which are towards the nose. Since rays passing
through a prism are bent towards its base (p. 22),
the rays from a point distant, say $8''$, after traversing two prisms with bases towards each other will
be less divergent, and on entering the eyes will
be referred to the point from which they would have
come, had they originally held their present (altered)
direction. To this point the eyes will be directed,
and as it lies further off than $8''$, less convergence
will be needed. The same result can be partly
gained by putting ordinary convex spectacle lenses
so near together that the patient looks through the
outer part of the glass, which then acts as a prism
with its base towards the nose.

CHAPTER XX

STRABISMUS AND OCULAR PARALYSIS

STRABISMUS exists whenever the two eyes are not
(as they ought to be) directed towards the same object.
The eye is "directed towards" an object when the
image is formed on the most sensitive part of the retina (the yellow spot) ; the straight line joining the
centre of this image with the centre of the object is
the "visual line," a prolongation of the "visual axis"
(compare p. 250). In health the action of the ocular
muscles is such as to keep both visual lines always
directed to the object under regard, and binocular but
single vision is the result. Although each eye receives
its own image, only one object is perceived by the sensorium, because the images are formed on parts of the
retinæ which "correspond" or are "identical" in
function, *i. e.* which are so placed that they always
receive identical and simultaneous stimuli.

But if, owing to faulty action of one or more of the
muscles, one eye deviates and the visual lines cease
to be directed towards the same object, the image
will no longer be formed on the yellow spot in both
eyes. In one of them it must fall on some other and
non-identical part of the retina, and the result will
be that two images of the same object will be seen.
In Fig. 63 *y* is the yellow spot in each eye, and the
visual line of the R. eye (the thick dotted line)
deviates inwards ; hence the image of the object, *ob.*,
which is formed at *y* in the L. eye, will in the R.
eye fall on a non-identical part to the inner side of *y*.
Ob. will be seen in its true position by the L. eye.
To the R. eye, however, it will appear to be at *F. ob.*,

because the part of the R. retina which now receives
the image of *ob.* was accustomed, when the eye was
normally directed, to receive images from objects in

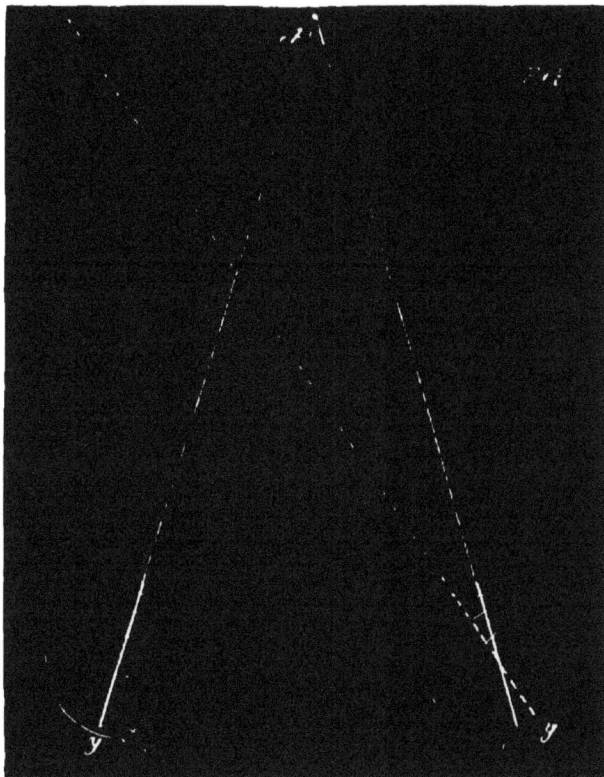

FIG. 63.—Shows the position of the double images in diplopia
from convergent or crossed strabismus.   The images are
homonymous, or correspond in position to the eyes.

the position of *F. ob.*; and in consequence of this
early habit *F. ob.* is the position to which every
image formed on this part of the retina is referred.
Since the image of *ob.* in the squinting eye is formed
on an eccentric part of the retina more or less distant from the central and most perfect part (the
y. s.), it will not appear so clear nor so bright as the
image formed in the sound eye; it is called the
"false" image, that formed in the sound eye being
the "true" one. The greater the deviation of the

visual line (*i. e.* the greater the squint), the wider
apart will the two images appear and the less distinct will the "false" image be.

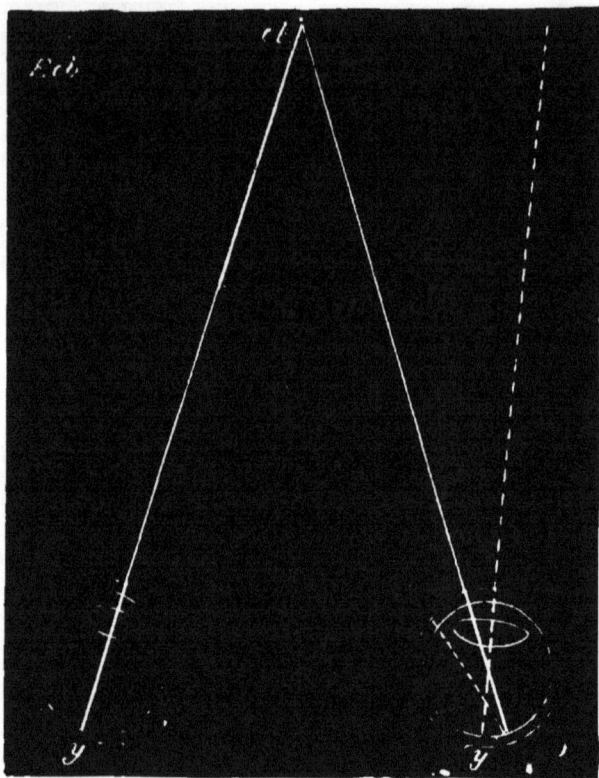

FIG. 64.—Position of double images in divergent strabismus.
The images are crossed.

[The y. s. (*y*) of the right eye will receive an image
of some different object lying in its visual line
(shown by the thick dotted line); this image, if
sufficiently marked to attract attention, will be seen,
and will appear to lie upon the image of *ob.* seen by
the sound (left) eye. Two equally clear objects will
be seen superimposed; but, as a rule, only one of
them is attended to, the perception of the other
being habitually suppressed or neglected (p. 230),
and the neglected image always belongs to the
squinting eye.]

If the eye deviate towards its fellow convergent squint, (Fig. 63, where the R. eye deviates towards the L.), the object of attention will seem to the squinting eye to be in the opposite direction; the image belonging to the R. eye being referred to the patient's R., and that belonging to the L. eye to his L. Hence in convergent or crossed strabismus the double images correspond in position to the eyes, or are *homonymous*. Similar reasoning will show that if the eye deviate from its fellow (as in Fig. 64, divergent squint), the position of the double images must be reversed, and the image belonging to the R. eye appear to be to the left of the other. Hence in divergent squint the double images are *crossed*.

Squinting is not always accompanied by double vision. (1.) If the deviation is extreme, the false image is formed on a very peripheral part of the retina, and is so dim as not to be noticed. (2.) As already mentioned, after a time the consciousness both of the "false image," and of the image formed on the y. s. of the squinting eye, is suppressed in order to avoid confusion. The result usually is that the eye, or rather its brain-centre, becomes permanently amblyopic, and in time almost blind. It must be stated, however, that this, the usually received, explanation of the amblyopia so often observed in eyes which have been squinting for a long time, has been assumed on the theory of congenital (rather than acquired) "correspondence" between the retinæ, and that it is doubted by some high authorities.

For the method of examining for strabismus and diplopia (see pp. 7 and 8).

Strabismus may arise from any one of the following muscular conditions—(1) over-action; (2) weakness following over-use; (3) disuse in an eye whose sight is defective; (4) stretching and weakening of the tendon after tenotomy; (5) from paralysis of one or more muscles.

(1.) Overaction of the internal recti gives rise to

the concomitant squint of hypermetropia (p. 265). Occasionally convergent squint occurs in high degrees of myopia. In both forms the external recti retain full power and nearly full range of movement.

(2.) Strabismus from weakness following overuse (muscular asthenopia, pp. 253 and 188) always depends no weakening of the internal rectus, and is consequently divergent. It is commonest in M., but is also occasionally seen in H., and in a less marked degree, without any error of refraction. The eye can usually be moved into the inner canthus, even in extreme cases, by making the patient look sideways, though not by efforts at convergence; and it is thus but rarely that these cases simulate paralysis. "Advancement" of the weakened muscle and tenotomy of its antagonist are often useful.

(3.) Strabismus from disuse is also nearly always divergent, depending, as it does, on relaxation of the internal rectus. It occurs in cases where convergence is no longer of service, as when one eye is blind from opacity of the cornea or other cause, or where the refraction of the two eyes is very different (p. 231). Treatment is seldom useful, but tenotomy of the external rectus may be called for.

(4.) Stretching and weakening of the internal rectus after division of its tendon for convergent squint, may give rise to divergent squint simulating paralysis of the internal rectus. The caruncle in these cases is generally much retracted, and this, together with the history of a former operation, will prevent any mistake in diagnosis. The squint can always be lessened, and often quite removed, by an operation for "readjustment" or "advancement" of the defective muscle.

(5.) *Paralytic squint.*—The deviation is caused by the unopposed action of the sound muscles. When the palsied muscle tries to act the eye fails, in proportion to the weakness, to move in the required direction. In many cases there is only slight paresis, and the resulting deviation is too little to be objectively noticeable; but in such cases the diplopia is very troublesome, and it is for this symptom that the patient comes under care (p. 8). Further, in these slight cases the symptoms often vary in a marked degree, according to the effort made by the patient. In paralysis of the third nerve the several branches are often affected in different degrees, and the resulting strabismus and diplopia are then complex. When paralysis is of long standing secondary contraction of the opponent seems sometimes to occur, still further complicating the symptoms. Lastly, the sound yoke-fellow * of the paralysed muscle sometimes acts too much in obedience to efforts made by the latter, and in this way the squint may, even when both eyes are uncovered, affect the *sound* instead of the paralysed eye; but such cases are rare (compare Secondary Squint, p. 8).

The commonest forms of paralytic squint are due to affection, separately, of the external rectus (sixth nerve), superior oblique (fourth nerve), or of one or all of the muscles supplied by the third nerve (internal, superior, and inferior recti, inferior oblique, levator palpebræ).

Paralysis of the external rectus (sixth nerve) causes a convergent squint from preponderance of the internal rectus, which, except in the slightest cases, is very noticeable. Movement straight outwards is impaired, and if the paralysis be complete the eye cannot be moved outwards beyond the middle line of the palpebral fissure. There is homonymous diplopia (p. 285); the two images, when in the horizontal plane, are upright and on the same level; the distance between them increases as the object is moved towards the paralysed side, but it diminishes, or the images even

* Yoked or conjugate muscles are the muscles of opposite eyes which act together in producing lateral and vertical movements; *e.g.* the internal rectus of one eye acts with the external rectus of the other in movement of the eyes to the R. or L.

coalesce, in the opposite direction.   Thus, in paralysis
of the left external rectus (Fig. 65, the uppermost
figure), the images separate more as the object is
moved to the patient's left, but approach one another,
and finally coalesce as it is moved over to his right.
In slight cases the diplopia ceases during convergence
for a near object, but reappears when gazing straight-
forwards at a distant object.   In the upper part of the
field the false image is also lower than the true one,
and in the lower part of the field it is higher.

In *paralysis of the superior oblique (fourth nerve)*
there is often no visible squint, or a slight convergent
and upward deviation may be noticeable.   But when
the eyes are directed below the horizontal, very troublesome diplopia arises from the defective downward
and outward movement, and loss of rotation of the
vertical meridian inwards, to which the lesion gives
rise.   In downward movements, especially downwards and towards the paralysed side, the eye remains
a little higher than its fellow; in trying to look
straight down (inferior rectus and superior oblique)
the unopposed action of the inferior rectus carries the
cornea somewhat inwards (convergent squint), and at
the same time rotates the vertical axis outwards,
whilst the cornea remains on a rather higher level
than its fellow ; in following an object from the horizontal middle line down-outwards it will be seen that
the vertical meridian of the cornea does not, as it
should, rotate inwards.

In many cases, however, the slight defects of movement caused by paralysis of the superior oblique are
not clearly marked, and the diagnosis has to be based
on the characters of the diplopia (compare p. 8).   In
all positions below the horizontal line the false image
will be below the true one, and displaced towards the
paralysed side (homonymous) ; thus, if the R. muscle
be at fault the false image will be below and to the
patient's R. (Fig. 65, arrow-head figure) ; further, it
will not be upright, but will lean towards the true

image. The difference in height between the images is greatest in movements towards the sound side; the lateral separation is greater the further the object is moved downwards; the leaning of the false image is greatest in movements towards the paralysed side. When the patient looks on the floor, *i. e.* projects the images on to a horizontal surface, the false image seems nearer to him than the true one. The images are always near enough together to cause inconvenience, and as the diplopia is confined to, or is worst in, the lower half of the field, the half most used in daily life, paralysis of the superior oblique is very annoying, especially in going up or down

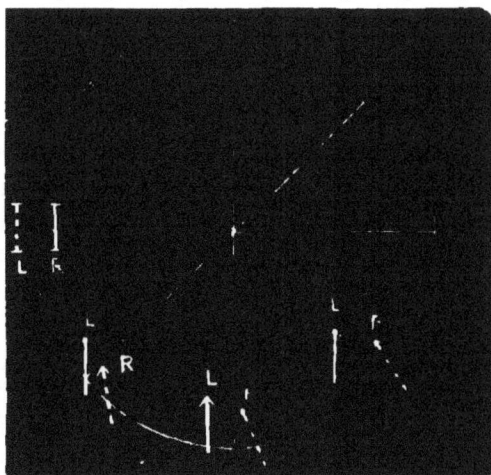

Fig. 65.—Chart showing position of double images, as seen by the patient, in paralysis of L. external rectus and R. superior oblique.

stairs, in looking at the floor, counting money, and similar acts.

Paralysis of the third nerve when complete, causes ptosis, loss of inward, upward, and downward movements, loss of accommodation, and partial mydriasis. There is well-marked divergent strabismus from unopposed action of the external rectus. The slight downward and outward movement effected by the superior oblique remains. The diplopia is crossed

· 19

(p. 285).   The mydriasis is never so great as that
produced by atropine.   But, in the majority of
cases, paralysis of the third is incomplete, affecting
some branches (and muscles) more than others, and
the result is a less typical condition than the above.
Complete isolated paralysis of a single third nerve
muscle is very rare.

PECULIARITIES OF PARALYTIC STRABISMUS

(1.) If the patient look at an object (say a pen
handle held at the ordinary distance of distinct
vision), and the sound eye be then covered by holding a card (or better, a piece of ground glass) before
it, the paralysed eye will make an attempt (more
or less successful according to the degree of the
palsy) to look at the object.   The absolute movement effected will call for a greater effort than if the
nerve were healthy, and as the eye muscles always
work in pairs, the same effort will be transmitted to
the companion muscle of the healthy eye, which
will, in consequence, describe a larger absolute movement than the paralysed eye, i.e. the secondary squint
will be greater than the primary (p. 8).   This test
is sometimes of use in distinguishing which is the
faulty eye, in cases where the squint is slight and
the patient unable to distinguish between the false
and true images (p. 9).   (2.) Giddiness when the
sound eye is closed.   This symptom varies much in
severity in different cases.   It depends on the erro-
neous judgment of the position of objects, which is
the result of the weakened muscle not being able to
achieve a movement of the eye, corresponding in
magnitude to the effort which it makes.   It is absent
when both eyes are open, and when the paralysed
eye is covered.   Like the former test it sometimes
helps us in doubtful cases to determine which is the
faulty eye.   ·
    Paralysis of the ocular muscles is seldom symme-
trical; in the rare cases where it is so, the disease is

always intracranial, and probably in most cases
nuclear, though symmetrical disease of nerve trunks
has been found in some cases. In certain cases of
symmetrical paralysis of all the ocular muscles (ophthalmoplegia externa), which are believed to depend
on nuclear disease, other cranial nerves (especially the optic and fifth) are also often involved, and
spinal symptoms often present.

PARALYSIS OF THE INTERNAL MUSCLES OF THE EYEBALL

The three internal muscles are supplied by two
nerves; the ciliary muscle and sphincter of the pupil
by the third nerve (short root of lenticular ganglion), the dilator of the pupil by the sympathetic,
but whether from the lenticular ganglion or by
branches independent of that structure is uncertain.
The following paralytic states of these three muscles
are to be distinguished.

A. *Iris affected alone.*—(1.) Paralysis of the dilator. The pupil in ordinary light is equal to or rather
smaller than the other; in a bright light it contracts
a little, but when shaded does not dilate beyond the
size it had in full light, and hence, if the eyes be
examined in a dull light, the paralysed pupil will be
much smaller than its fellow (paralytic myosis);
accommodation is not affected. The condition is
seen as a part of paralysis of the cervical sympathetic,
and probably under other conditions; in a certain
degree it is common, perhaps natural, in old age.
(2.) Paralysis of the sphincter alone (paralytic mydriasis) causes moderate dilatation; the pupil remains
of the same size in the brightest light, and accommodation is unaffected. It is very rare. (3.) Paralysis of both iridal muscles without affection of
accommodation (iridoplegia). The pupil is of medium
size and quite uninfluenced by variations of light.
But in many, perhaps in all, cases it still acts in
association with the ciliary muscle, becoming, as in

health (p. 15), smaller when a near object is looked at, and larger when the patient gazes into the distance, even though the distant object (*e. g.* the sky) be much brighter than the near one.

B. *The ciliary muscle paralysed alone* (Cycloplegia).—Accommodation is lost without any change in the activity of the pupil. The term is applied only to cases of nervous origin, not to presbyopia. The condition is very rare except after diphtheria, when paralysis (often only paresis) of accommodation, with little or no affection of iris, is common.

C. *Ciliary muscle and iris both affected.* (1.) Mydriasis and cycloplegia; partial dilatation of the pupil (to about four or five mm.), with loss of accommodation. This is the common condition in complete paralysis of the third nerve, and in rare cases it is seen without failure of any other part of the nerve. (2.) Paralysis of all the three internal muscles ("ophthalmoplegia interna," Hutchinson); loss of accommodation with absolute immobility of the iris to light, the pupil being of about medium size.

Causes of ocular paralysis.—It is convenient to separate the external and mixed forms, from those in which only the internal muscles are involved, since the local causes are, as a rule, different in the two groups.

Paralysis of the third, fourth, or sixth nerve may be the result of tumours or other growths in the orbit, but in such cases, as a rule, the paralysis forms only one amongst other well-marked local symptoms. In the vast majority of uncomplicated ocular palsies there is nothing in the state either of the eye or the orbital parts, to guide us in determining whether the disease is seated in the orbit or within the cranium. Meningitis, morbid growths, and syphilitic periostitis at the base of the skull or involving the sphenoidal fissure, and aneurism of the internal carotid in the cavernous sinus, often cause ocular palsy, usually not confined to a single nerve.

Syphilitic gumma of the nerve trunk is probably the commonest cause of single paralysis ; the intracranial portion of the nerves is known to be often the seat of such growths, but neural gummata probably occur also on the orbital part of the nerves where they are too small to cause proptosis or signs of inflammation. Fractures of the skull often lead to ocular paralysis by compression of a nerve, either by displacement of bone or by inflammatory exudation afterwards thrown out. Paralysis of the third nerve, coming on simultaneously with hemiplegia of the opposite side, may indicate a lesion in the crus cerebri on the side of the paralysed third. In certain cases there are neither symptoms nor facts enabling us to locate the seat or prove the cause of the paralysis. The term "rheumatic" is often applied to these cases on the assumption that the palsy is peripheral and caused by cold, that it is, in fact, to be compared to peripheral paralysis of the facial nerve ; no doubt some of them are in reality syphilitic cases. Paralysis, usually of short duration and affecting only one nerve, is not uncommon at an early stage of locomotor ataxy. Ophthalmoplegia externa is believed by Mr Hutchinson to be indicative of disease of the nerve centres, and to be very frequently caused by syphilis.

In respect to the causation of the internal paralysis we have but little positive knowledge. Paralysis of the three internal muscles without any affection of the external muscles (ophthalmoplegia interna) is believed by Mr Hutchinson to indicate disease of the lenticular ganglion. Mydriasis, with cycloplegia and no other paralysis, could be best accounted for by the supposition of disease of the short (third nerve) root of the lenticular ganglion. Iridoplegia without cycloplegia is difficult to explain; but it has been suggested that in these cases, where the iris is paralysed to light but acts with accommodation, there is an interruption in the nerve circuit between the

optic and third nerve-centres, the centres themselves
and their peripheral branches being intact. This
hypothesis, however, does not appear to explain the
cases, which are by no means rare, in which a paralysis, at first limited to the iris and related only to
light and shade, later on affects the ciliary muscle
and the accommodation-action of the iris; nor for
the fact that when such cases recover the action of
the iris to light is the last to return.

Treatment.—In estimating the results of treatment
it is well to remember, first, that some cases of ocular
palsy recover spontaneously, and, next, that in many
the defect is a paresis rather than paralysis, and
that in such the symptoms often vary in severity
from day to day or even whilst under observation at
a single visit, according as the patient devotes more
or less attention and effort to the test movements
called for. The questions of syphilis and of injury
to the head are always to be carefully inquired into,
especially when only one nerve is paralysed. When
several nerves are involved, tumour, aneurism, or
syphilis (either gummatous inflammation at the
base, or nuclear disease) are to be suspected; in the
last-named affection there is usually bilateral symmetry. Iodide of potassium and mercury are the
only internal remedies likely to be beneficial, and
unless syphilis be quite out of the question they
should have a full trial; many cases recover quickly
under moderate doses of iodide. Localised electri-
sation of the paralysed muscles is valuable in some
cases.

Nystagmus (involuntary oscillating movements of
the eyes) is generally associated with serious defect
of sight dating from very early life, such as opacity
of cornea after ophthalmia neonatorum, congenital
cataract, and choroido-retinitis. It is, however, also
seen in cases of infantile amblyopia without apparent
cause, and in albinoes. Nystagmus is occasionally

·leveloped during adult life, in miners who work in cramped positions, looking upwards, and with a bad light; and it forms a symptom in some cases of disseminated sclerosis.

In most cases both eyes are affected, but when the defect of sight is monocular, the nystagmus is sometimes unsymmetrical. The movements in nystagmus vary much in rapidity, amplitude, and direction in different cases, and even in the same case at different times; they are generally worse when the patient is frightened or nervous. In many cases the nystagmus becomes much less marked as life advances.   No treatment is of any use. ,

CHAPTER XX

OPERATIONS

A. OPERATIONS ON THE EYELIDS

1. *Epilation* in ophthalmia tarsi.—Position: patient seated; surgeon standing behind. The forceps used should be broad-ended, with smooth or very finely roughened blades which meet accurately in their whole width. Stretch the lid tightly by a finger placed over each end. Pull out the lashes at first quickly in bundles, and finish by carefully picking out the separate ones that are left.

2. *Eversion of upper lid.*—Position as for 1, or the surgeon may stand in front. The patient looks down, a probe is laid along the lid above the upper edge of the cartilage, and the lashes, or the edge of the lid, seized by a finger and thumb of the other hand, and turned up over the probe,which is simultaneously pushed down. After a little practice the probe can be dispensed with, and the lid everted by the forefinger and thumb of one hand alone, one serving to fix and depress the lid, the other to turn it upwards.

3. *Removal of meibomian cyst.*—Position as for 1. Instruments: a small scalpel or a Beer's knife, such as was formerly used for extraction of cataract (Fig. 88), and a curette, or equivalent instrument (Fig. 85). (1) Evert the lid; (2) make a free crucial incision into the tumour from the conjunctival surface; (3) remove the growth either by squeezing the lid between finger- and thumb-nail, or by freely stirring up the cavity with the curette. After treatment, cold bathing. The cavity fills with blood, and may thus for a few days be larger than before. These tumours have no distinct cyst-wall.

4. *Inspection of cornea* in purulent ophthalmia. Position : if the patient be a baby or child, the back of its head is to be held between the surgeon's knees, its body and legs being on the nurse's lap ; if an adult, the same as for 1. If the lids cannot be easily separated enough by a finger of each hand to allow a view of the cornea, one or both fingers should be replaced by some form of elevator (a convenient pattern is shown in Fig. 66), by which

FIG. 66.—Desmarres' lid elevator.

the lid can be raised and held away from the globe. If this instrument be gently used we avoid all risk of causing perforation of the cornea should a deep ulcer be present, an accident which may happen in cases attended by much swelling or spasm of the lids if the fingers are used.

5. *Entropion.—Spasmodic entropion of the lower lid,* with relaxed skin, in old people. Position as for 1.

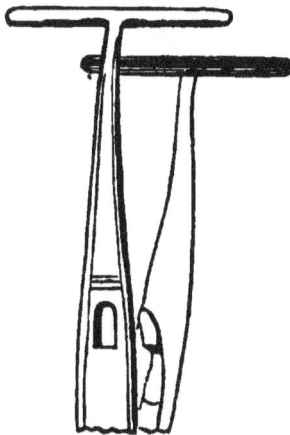

FIG. 67.—Entropion forceps.

Instruments :—T forceps (Fig. 67), straight strabismus scissors (Fig. 73), toothed forceps. (1.) With,
the T forceps pinch up a fold of skin as close as possible to the edge of the lid and of width proportioned
to the degree of inversion, and cut it off close to the
forceps ; (2) with the toothed forceps pinch up the
orbicularis muscle now exposed, and cut out a small
piece.   Sutures need not be used.

    6. *Organic entropion and trichiasis.* —(1.) When the
whole row of lashes is turned inwards, and the inner
surface of the lid much shortened by scarring, the
radical extirpation of all the lashes is the quickest

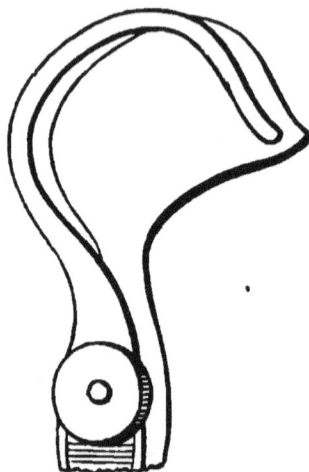

FIG. 68.—Snellen's lid clamp (for the R. upper lid).

and most certain means of giving permanent relief,
but it leaves, of course, an unsightly baldness.   Position : recumbent ; the surgeon stands behind the
patient.   Anæsthesia seldom necessary.   Instruments : a horn or bone lid-spatula, or a lid clamp
(Fig. 68 shows the clamp for the right eye), a
small scalpel or Beer's knife, and forceps.   Make
an incision from end to end (leaving the punctum) between the hair-follicles and meibomian
ducts; as if about to split the lid into two layers.
Then make a second incision through the skin and

tissues, about one twelfth of an inch beyond the border of the lid in a plane at right angles to the former incision. The strip of skin and cartilage included between these two cuts will now be almost free, except at its ends, which are to be united by a cross cut, and the strip dissected off. The strip removed should include the hair-follicles in their whole depth. Lastly, examine the white edge of the cartilage, now exposed, for any hair-follicles accidentally left behind; they will appear as black dots, which are to be carefully removed, or they may produce fresh hairs.

In the same or slighter cases, the inversion of the border of the lid may be much lessened by division of the scar from the conjunctival surface (Burow's operation). The wound gapes, and the inverted border of the lid falls forwards and is kept in its natural place by the cornea. The only instrument needed is a scalpel and no anæsthetic is necessary. Position as for 1, or recumbent. The lid is kept well everted whilst the incision is being made, and the knife held *perpendicular* to the surface of the lid, and parallel to its border; the whole thickness of the cartilage is to be divided from end to end, the test being the appearance of a bluish hue seen through the skin on replacing the lid. This operation gives complete relief for the time, but may need repetition in a few months.

Various operations are performed for transplantation of the displaced lashes forwards and upwards, so as to restore their natural direction. *Arlt's operation.*—The free border of the lid is split from end to end (leaving the punctum) as for extirpation of the lashes, but more deeply. A second straight incision is now made through the skin parallel to and about two lines from the border of the lid, down to, but not through, the cartilage; thirdly, a semilunar incision is made, joining the last at each end and including, therefore, a semicircular flap of

skin, of greater or less width, according to the effect
desired. This flap is now dissected off without
injury to the orbicularis, and the edges of the
wound are brought together with sutures. The anterior layer of the lid-border, which contains the
lashes, is thus tilted forwards and drawn upwards.

If a greater effect is wanted, a wedge-shaped strip
of the tarsal cartilage may be removed parallel with
and about a line from the border of the top lid, by
cutting through or separating the fibres of the orbicularis after the skin flap has been removed. A wedge-
shaped groove is thus made in the tarsus, and more complete eversion of the border is gained (Soelberg Wells'
combination of Arlt's and Streatfeild's operations).

A third operation, that of Streatfeild, consists in the
removal of the wedge-shaped strip (with its superjacent skin and muscle) at a distance of about a line
or two from the lid border, and without splitting the
cartilage. No sutures are used.

7. *Ptosis* (chiefly the congenital form) may be treated
by the removal of an oval of skin from the upper lid
parallel to its length, the muscle not being touched.
Sutures are to be carefully inserted, and every effort
made to get immediate union, so as to avoid a scar.

8. *Division of outer canthus* for spasm of lids in
children or to diminish pressure on the eye in severe
purulent ophthalmia. Position as for inspection of
cornea. The division is made down to the bony rim
of the orbit by one bold stroke of a strong scissors,
or with a bistoury or scalpel. There is free bleeding
for a few minutes. The wound heals in a few days
unless purposely kept open.

9. *Canthoplasty.*—An operation for lengthening
the palpebral fissure at the outer canthus. The
canthus is divided, as in 8, but with more care as to
smoothness of incision, a bistoury and director being
used. The contiguous ocular conjunctiva is then
attached by sutures to the cut edges of the skin so as

to prevent reunion, one suture being placed in the angle of the wound, one above, and one below.

B. OPERATIONS ON THE LACHRYMAL APPARATUS

1. *Lachrymal abscess* (suppuration in and around the lachrymal sac) is generally opened through the skin over the lower part of the sac where it points, the line of incision being down and out. Some prefer to open it from the canaliculus.

2. *Abscess or mucocele of the frontal sinus.* —Make a free incision into the swelling through the skin with a scalpel. Explore the cavity with the finger; then passing a finger up the corresponding nostril introduce a trochar, into the opening made by the scalpel, and with it perforate the floor of the cavity on to the finger in the nostril. Now pass a drainage tube or thick seton along the same course by means of a probe, bring it out at the nostril and tie the two ends together. It is to be worn for several weeks at least.

3. *Slitting up the lower canaliculus.*—This is done either by running a Beer's knife (Fig. 88) or a small scalpel along a fine grooved director (Fig. 69) introduced into the whole length of the canaliculus; or by a knife with a blunt or probe point, and a blade narrow enough to enter the punctum. The best forms of these knives are Weber's knife with a probe end, Bowman's, shaped like a miniature dinner-knife, with nearly parallel borders and a rounded end (Fig. 70) and Liebreich's (Fig. 71). Position as for 1. (1.) The lower lid is drawn tightly outwards and everted a little; this is done by the thumb if a canaliculus knife is used, by the third or ring finger if the director is used,

FIG. 69.—Canaliculus director.

the thumb and index holding the director when
in position. (2.) The director or canaliculus knife
is passed *vertically* into the punctum, and then
turned horizontally and passed on through the neck
of the canaliculus till it reaches the bony (inner) wall
of the lachrymal sac. (3.) If the director is used,
the Beer's knife is now slid along it quite to the inner
wall of the sac, thus making sure that the whole length

FIG. 70.—Bowman's canaliculus knife.

FIG. 71.—Liebreich's knife for canaliculus and nasal duct.

of the canaliculus is divided. If Weber's or Bowman's
knife be used, it is pushed on into the sac in the same
way as the director, and then raised up from heel
towards point, and thus made to divide the canaliculus, care being taken that the neck is freely divided.
Liebreich's knife cuts its own way, like Beer's cataract knife, without being raised. In cases of mucocele it is best to divide the wall of the sac very freely.
Some surgeons divide the upper canaliculus.

4. *Catheterism of the nasal duct.*—After dividing
the canaliculus pass a No. 6 Bowman's lachrymal
probe horizontally along its floor into the sac till it
strikes the inner bony wall. Then raise it to the
vertical and push it down the duct (downwards and
a little outwards and backwards) till the floor of the
nose is reached. This probe is $\frac{1}{20}$th inch in diameter
and is the largest of Bowman's set of six. Try smaller
numbers if an impassable stricture be found. Larger
probes, and of various patterns, are often used.

5. *A stricture of the duct* may be incised with any of
the canaliculus knives, although Weber's and
Bowman's are too slender to be used with safety.

Liebreich's is intended to be so used, and a special knife for the purpose had previously been introduced by Stilling. The knife in any case is used as a probe, being pushed quite down the duct, then partly withdrawn and turned in other directions, and pushed down again. There is generally bleeding from the nose.

In all these procedures it is essential to be certain that the probe or knife rests against the bony (nasal) wall of the lachrymal sac before it is turned into the vertical direction. If the probe be stopped at the entrance of the canaliculus into the sac (as may easily happen if the canal be not thoroughly slit in its whole length), the lid will be pulled upon and puckered whenever the instrument is pushed towards the nose ; but if the instrument has reached the sac, backward and forward movements will not usually cause puckering of the lid. If in the former case the instrument be turned up, and an attempt made to pass it down the duct, a false passage will be made.

The direction of the nasal ducts is such that if prolonged upwards they would converge and meet in the middle line three or more inches above the sac. If the upper end of the probe takes a direction different from this (e.g. parallel to the middle line or diverging from it), a false passage is generally indicated. Slight variations in direction, however, are met with.

6. *Abscess of the lachrymal gland* or of the orbit is generally to be opened through the skin, though a conjunctival opening is sometimes satisfactory. The abscess generally points to the skin. All incisions should be parallel to the eyebrow. Drainage tube is usually needed.

The rarely needed operations for the removal of the lachrymal gland, and for the obliteration of the lachrymal sac for the cure of intractable mucocele, need not be described here.

C. OPERATIONS FOR STRABISMUS

Tenotomy.—The object is to divide the tendon subconjunctivally and close to its insertion into the

sclerotic. The internal and external recti are the only
tendons commonly divided, and the internal by far the
morefrequently. Anæsthesia is seldom necessary except
for children. Position recumbent.—The operator
usually stands on the patient's right side for whichever
eye is to be operated on, but some prefer to stand
behind, and use curved scissors. Instruments: Stop
speculum (Fig. 72 shows a convenient and common pattern); scissors, most operators preferring them straight,

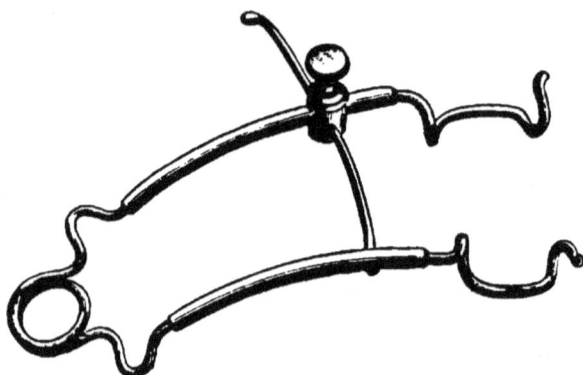

FIG. 72.—Stop spring speculum.

and more or less blunted at the points (Fig. 73);
toothed fixation forceps (Fig. 74); strabismus hook
(Fig. 75). There are several patterns of hooks, differing in the length and sharpness of the curve, and
in the form of the tip, some having it slightly bulbous,
others flattened but not enlarged at the end. I
prefer the hook to be flattened sideways.

Operations.—*Critchett's operation.*—Taking the fixation forceps in the left hand, pinch up a fold of conjunctiva over the lower border of the tendon at its
insertion (say of the right internal rectus). With
the scissors make a small opening in this fold close
to the forceps end, the cut being made in the direction of the caruncle. The capsule of Tenon is now
identified as a layer of fascia, which can be moved
over the sclerotic, and pinched up separately; in
this capsule an opening is to be made corresponding

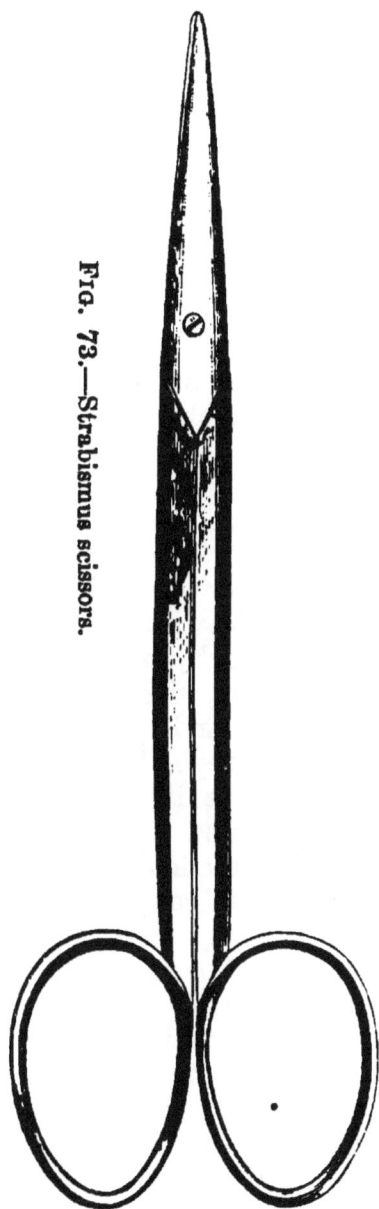

FIG. 73.—Strabismus scissors.

FIG. 74.—Fixation forceps; the teeth shown in section at the end.

FIG. 75.—Strabismus hook (the bent part is drawn rather too thin).

to the conjunctival wound. By taking deep hold with the forceps both conjunctiva and capsule may often be divided at one stroke, but with less certainty

20

than in separate stages. There are great differences in the thickness, both of conjunctiva and Tenon's capsule, in different persons.

Now take the hook in the right hand (retaining the lip of the wound with the forceps in the left), and pass it, concavity downwards and point backwards, through the opening in the capsule as far as its elbow, keeping its end always flat against the sclerotic. Next turn the end of the hook upwards by sweeping it, still guided by the sclerotic, between the tendon and the globe until its end is seen projecting beneath the conjunctiva above the upper border of the tendon. On now attempting to draw the hook towards the cornea it will be stopped by the tendon. If Tenon's capsule have not been well opened the hook will not sweep round, nor can it be passed beneath the tendon.

Next lay down the forceps, transfer the hook to the left hand, holding its handle parallel with the side of the nose, and tightening the tendon by traction forwards and outwards; pass the scissors, with the blades slightly opened, into the wound, and push them straight up between the hook and the eye. The tendon being included between the blades, is divided at two or three snips, with a crisp sound and feeling. When the whole breadth of the tendon is divided the hook slips forwards beneath the conjunctiva up to the edge of the cornea. It is always best to withdraw the hook partly or wholly, and by reintroducing it, to make sure that no small strands of the tendon have escaped division, for the operation does not succeed unless the division be quite complete.

No after-treatment is needed, but the patient is more comfortable if the eye be tied up for a few hours. If there be much conjunctival bleeding (as is common when no anæsthetic is used) a second small hole may be cut in the conjunctiva over the upper border of the tendon, to let the blood escape.

The effect of the tenotomy may, if necessary, be

increased by tying the eye out; a stout suture is passed through the conjunctiva, embracing about a quarter of an inch, close to the outer border of the cornea, and the eye being drawn outwards, the two ends of the thread are firmly attached by strapping to the skin of the temple. It may be left in for three or four days when it will usually have cut out.

The difficulties for beginners are—(1) to be sure of opening Tenon's capsule; (2) to avoid pushing the tendon in front of the scissors, especially when only the upper part remains undivided.

Division of one internal rectus by this operation diminishes the squint by about two lines.

The tendon, in retracting, draws with it, to a varying extent, the neighbouring parts of Tenon's capsule, and the superjacent conjunctiva. Two consequences follow:—1st. The tendon, owing to these indirect but wide attachments, cannot retract fully, and hence the maximum effect of its division is not obtained. 2nd. The caruncle is drawn back by the retreating tendon, and some deformity results; this, however, will be very slight if the operation wound be made small, and as near as possible to the cornea.

Liebreich's operation.—After making the conjunctival wound as above, the scissors are passed between the conjunctiva and Tenon's capsule, and by repeated horizontal snips are made to separate these membranes freely from one another over the tendon, as far as the caruncle. The capsule is then opened and the tendon divided as in the former operation. The conjunctival wound is closed by a suture. When well performed, this operation gives a greater effect with less retraction of the caruncle than Critchett's operation, but it is more tedious and painful than the latter. The eye may be tied out if we wish the maximum effect.

The immediate effect of the tenotomy of a rectus muscle is somewhat lessened after a few days by the reunion of the tendon with the sclerotic, but after a

few weeks or months it is again increased by the
stretching of this new tissue (final stage).

Readjustment or *Advancement* consists in bringing
forwards to a new attachment the tendon of a rectus
(generally the internal, occasionally the external),
which has become attached too far back after a previous tenotomy or has become weakened in myopia.
There are several different operations, but in all of
them the tendon is held in its new position by sutures.
The operation is tedious and painful, and the patient
must always be under an anæsthetic. The instruments are the same as for tenotomy.

The following operation (Critchett's) is a convenient and efficient one. (1) Make a vertical opening
in the conjunctiva in length equal to, or rather longer
than, the width of the tendon, and about one eighth
of an inch from the corneal border; make a short
horizontal cut from each end of the incision towards
the canthus, thus forming a short rectangular
flap. (2) Divide the subconjunctival fascia at the
lower border of the tendon, pass the hook under the
tendon in the usual way and bring its point out at
the upper edge of the wound. (3) Carry three fine
sutures (on curved needles by means of a needle
holder) through the conjunctiva and tendon, entering
them about a quarter of an inch to the canthal side
of the wound and bringing them out in the wound
between hook and sclerotic. By this means the
whole tendon is more certain to be included than if it
be divided before the sutures are passed; the inclusion of the conjunctiva and fascia in the sutures
gives a much better hold than if they be passed
through the muscle alone. (4) The assistant takes
the hook, and by it keeps the tendon on the stretch;
the operator takes hold of both ends of the sutures
in a bunch and drawing them tightly away from the
globe carefully snips the tendon from the sclerotic
without cutting the sutures; the hook which is thus
freed is now withdrawn. (5) The sutures are now

passed through the corneal flap of conjunctiva, care being taken to include as much tissue as possible for fear that the threads should cut out too soon. On now tightening up the sutures, the tendon (and its overlying parts) are brought forwards, and an idea can be formed of the effect that will be gained. It is generally necessary at this stage to shorten the flap containing the tendon, care being taken, however, to leave a good hold for the sutures. (6) Pass a stout suture through the conjunctiva close to the opposite edge of the cornea; by fastening this suture with strapping to the bridge of the nose, the eye will be drawn inwards and all tension taken off from the stitches of the readjusted tendon. Many operators divide the external rectus at this stage, but it is by no means always necessary, and sometimes is followed by disfiguring prominence of the eyeball; enough effect can often be obtained without it, whilst if necessary, it can be done afterwards. (7) While the assistant draws the eye straight inwards by the traction suture, the three wound sutures are tied, and, finally, the traction suture strapped to the nose and the eye lightly dressed with a wet bandage. In passing and tying the sutures, care is needed to ensure their all having an equal effect, so that the eye may be drawn straight inwards; the tendon may otherwise be implanted so as to draw the eye a little up or down as well as in, and an upward or downward squint and troublesome diplopia may result. The traction suture cuts out in three or four days; the tendon sutures should be left in a week. The pain and swelling, which for a few days are sometimes considerable, are best relieved by application of ice or a spirit lotion to the lids. The final result is not reached for several weeks.

D. EXCISION OF THE EYE

Instruments as for squint, but the scissors should be curved on the flat. The operator may stand either behind or in front. (1.) Divide the ocular conjunctiva all round close to the cornea, but leave, at one side, enough to hold by with the forceps. (2.) Open Tenon's capsule and divide each rectus tendon and the neigh-bouring fascia on the hook except the two obliques, which are seldom divided on the hook. (3.) Make the eye start forwards by pressing the speculum back-wards, so as to get behind the equator. (4.) Pass the scissors backwards along the side of the globe till their opened blades can be felt to embrace the optic nerve (recognised by its toughness and thickness) and divide it by a single cut while steadying the globe with a finger of the other hand. Finish by dividing the oblique muscles and remaining soft parts close to the globe. Apply pressure for a minute or two, and then tie up tightly with an elastic pad or small sponges overlaid by cotton wool for six or eight hours. There is scarcely ever serious bleeding. The artificial eye may be fitted in from two to three weeks.

After some weeks or months a button of granula-tion tissue sometimes grows from the scar at the bottom of the conjunctival sac, and should be snipped off.

The operation is more difficult when the eye is ruptured or shrunken, or the surrounding parts much inflamed and adherent. The order of division of the muscles is quite immaterial. The important points are to leave as much conjunctiva as possible, so as to form a deep bed for the glass eye, and by keeping the scissors close to the globe during the whole opera-tion, to avoid unnecessary laceration of the tissues.

When, as in some cases of intraocular tumour, it is desired to remove another piece of the optic nerve, the nerve should be felt for with the finger, seized and drawn forward with the forceps, and cut off further back with the scissors.

Abscission is the removal of a staphylomatous cornea with the front part of the sclerotic, leaving the hinder part of the globe with the muscles attached, to serve as a moveable stump for carrying the artificial eye.

Four or five semicircular needles carrying sutures are made to puncture and counter-puncture the sclerotic just in front of the attachments of the recti; the part of the globe in front of the needles is then cut off, the needles drawn through, and the sutures tied. The operation is admissible only in cases when the ciliary region is free from disease, the eye being lost by the results of corneal disease only. Such cases are not frequent, and the operation has, therefore, a very limited application; even in the most favorable cases the stump is not entirely free from the risk of setting up sympathetic inflammation. Modifications have been introduced in which the sutures are passed only through the conjunctiva or the muscles, and the risk is said then to be less than when they are passed through the sclerotic.

E. OPERATIONS ON THE CORNEA

Removal of foreign bodies from the cornea.—Position as for 1. Instruments: a steel or platinum spud (Fig. 76) or a broad needle with double cutting edge (Fig. 77). The eyelids are held open by the

FIG. 76.—Corneal spud.   FIG. 77.—Broad needle.

index and ring fingers, and the eyeball steadied by the intermediate finger placed against the temporal side of the globe. The chip of metal or stone is gently picked or tilted off by placing the edge of the spud beneath it, or, if firmly embedded, a certain amount of scraping may be necessary. The first few touches, by which the epithelium is removed, cause more pain than the after picking or scraping. When barely embedded in the epithelium, a touch with a little roll of blotting-paper is sometimes enough to detach the chip. When a fragment of iron has been present for more than a couple of days, its corneal bed is usually stained by rust, and a little plate or ring of brown corneal slough can often be picked off after the removal of the chip; but, as a rule, this minute slough will separate spontaneously.

When a splinter is deeply and firmly imbedded, especially if it have penetrated the cornea and be projecting into the anterior chamber, the operation is much more difficult, and is in fact no longer a "minor" one. Unless great care be taken the splinter in such a case may be pushed on into the chamber, and the iris or lens be wounded. This may sometimes be prevented by passing a broad needle through the cornea at another part and laying it against the inner aspect of the perforation, so as to form a guard or foil to the foreign body, the latter being removed by spud or forceps from the front.

After-treatment.—The protection of the corneal surface from friction and irritation by keeping the eye tied up is generally sufficient; a drop or two of castor oil placed in the conjunctival sac lubricates the cornea and lessens the irritation. Atropine is to be used if iritis be threatened.

A foreign body on the iris should, if recent, always be removed. In old cases we may judge by the symptoms whether to operate or not. The piece of iris on which it lies must generally be excised.

Paracentesis of the anterior chamber.—Position as for 1 or recumbent; anæsthesia seldom necessary. Instruments: a paracentesis needle (Fig. 78) with a very small, short, triangular blade bent at an

FIG. 78.—Paracentesis needle and probe mounted on same handle.

obtuse angle (like a minute bent keratome), or a broad needle (Fig. 77).

The former is the safer, as the blade is too short

to reach the iris or lens, even if the patient should jerk his head. One lip of the wound is depressed by a fine probe (Fig. 78), or similar instrument to allow the contents of the chamber to escape. In cases where the operation needs repetition every day or two the original wound can be reopened with the fine probe, but if several days elapse a fresh puncture is necessary. Speculum and fixation forceps should be used unless the patient has good self-control.

Corneal section for Hypopyon ulcer.—Position recumbent. Anæsthesia not usually needed. Instruments: a Graefe's or Beer's cataract knife (Figs. 83 and 88), speculum and fixation forceps. The incision is carried through the whole thickness of the cornea from one side of the ulcer to the other, being both begun and finished in sound tissue; or it may be placed entirely in sound cornea leaving the ulcer untouched; the object being to make a free opening into the anterior chamber without prolapse of the iris.

The knife is entered at an angle with the plane of the iris and with its edge straight forwards; when its point is seen or judged to have perforated the cornea, the handle is depressed until the back of the knife lies parallel with the iris, and the blade then pushed straight across the ulcer to the point chosen for counter puncture, or more often in practice is just pushed on till it cuts its way out. The aqueous ought not to escape until the point of the knife is engaged in its counter puncture, but an earlier escape cannot always be avoided. Notwithstanding the apparent risk to the iris and lens, accidents very seldom happen if the back of the knife is carefully kept parallel to them, or the point even directed a little forwards. Of course the incision is considerably shorter at its inner than at its outer surface. If it is desired to keep the wound open, its edges are to be separated by a probe every second or third day. The wound closes quickly at first, unless kept open, but after having been opened a few times, it sometimes remains patent for longer.

Operations for conical cornea.—The object is to produce a scar at the apex of the protrusion, which by its contraction shall reduce the curvature of the cornea, and so diminish the high degree of irregular myopic astigmatism to which the condition gives rise. Sight is often finally much improved, even though the scar remains nebulous, and an artificial pupil may be needed before the full advantage is gained.

There are two methods. Graefe's operation consists in first carefully shaving off the apex of the cone without entering the anterior chamber, and then applying solid mitigated nitrate of silver to the raw surface, the resulting ulceration being followed by some scarring. The application needs great care, and the after-treatment is troublesome, as there is the risk that more inflammation than is wished for may set in.

In another operation the apex of the cone is cut off with a cataract knife, the anterior chamber being entered, and the wound either left to close or united by sutures. There are several different modes of removing the little piece of cornea.

After-treatment.—Atropine and compressive bandage until the wound has closed; antiphlogistic treatment should inflammatory symptoms arise.

All operations for conical cornea are difficult and somewhat uncertain in result, but in many cases very great improvement is gained provided the vision was very bad at the time of operation. An artificial pupil is often necessary after a sufficient interval, if the corneal opacity be too large.

F. OPERATIONS ON THE IRIS

A portion of the iris is very often removed by operation (iridectomy), and with various objects. The principal of these are—(1) the direct improvement of sight by altering the position and size of the

pupil (artificial pupil); (2) to influence the course of
an active disease—glaucoma, iritis, ulcer of cornea
with hypopyon; (3) to remove the risks attending
exclusion and occlusion of the pupil, by restoring
communication between the anterior and posterior
chambers; (4) as a stage in the extraction of cataract.

Artificial pupil.—The object is to remove the portion of iris in the position best adapted to sight, thus
in cases of leucoma the iridectomy is made opposite the
clearest part of the cornea. When the state of the cornea allows it, the new pupil should be made down and
inwards or straight downwards; the next best place is
outward or out-upward, and straight upwards is, of
course, least useful, because the new pupil will be
covered by the lid. The iridectomy should generally
be small, and in many cases only the inner (pupillary) part of the chosen portion is to be removed, the
outer (ciliary) part being left so as to prevent the
light from passing through the margin of the lens.
After such an operation the pupil will be oval or
pear shaped, and widest towards the centre. The
incision should lie in the corneal tissue, if only the
pupillary part of the iris is to be removed; but if
only a narrow zone of cornea remains clear the incision must lie a little outside the sclero-corneal
junction, lest its scar should interfere with the transparency of the remaining clear cornea. The little
loop of iris should be cut off with a single snip.

In glaucoma the iridectomy is to be large (about
one fifth of the circumference); the iris must be
removed quite up to its ciliary attachment, and the
incision be made as far back in the sclerotic as
possible. The sides of the coloboma (the gap in the
iris) should be parallel, or even wider towards the
incision than towards the natural pupil. The loop of
iris, when drawn out, is usually cut first in one angle
of the wound, then torn from its ciliary attachment
by carefully drawing it over to the other angle of the
wound, and its other end then cut, the points of the

scissors being pushed just within the lips of the wound to ensure removal of the largest possible portion.

The difficulties in making an artificial pupil (for optical purposes) of the best shape, *i. e.* broad towards the natural pupil and narrow towards the circumference, are much greater than would be at first supposed, and several devices have been invented for ensuring this result.   In Mr Critchett's *iridodesis* the loop of iris is withdrawn through a small opening, and included in a fine ligature placed over the incision. The little bit shrivels and soon drops off, and the result is a pear-shaped pupil, with its broad end towards the centre. The inclusion of iris in the track of the wound has sometimes set up severe irritation, and even destructive irido-cyclitis, and on this account the operation is now but seldom performed.   Another plan is to draw out·a small loop of iris with a blunt silver or platinum hook, and to cut off only the pupillary portion. The objection to this method is its uncertainty; the hook may bring out too much or may tear or perforate the iris in the act of withdrawal; on the whole, however, it gives good results. Mr Carter cuts out a V-shaped bit of iris by introducing a small pair of blunt-ended scissors (Wecker's iridotomy scissors) into the anterior chamber, opening the blades and cutting out just as much as intrudes between them when the aqueous escapes. This operation requires much nicety, but when well performed it gives an excellent artificial pupil. The same iridotomy scissors may be used for simply cutting a notch in the iris, one blade being introduced with great care between the iris and the lens, and the other in front of the iris; the notch made in the iris by closing the blades will gape enough for the purpose required. The principal use of *iridotomy*, however, is in cases where extraction of cataract has been followed by severe iritis with thick and dense false membrane; in such cases a free incision, through the membrane or the iris, or both, will gape widely enough to give a good pupil and good sight.

The operation of iridectomy.—Position recumbent; the operator usually stands behind. Anæsthesia is always strongly advisable, though in urgent cases iridectomy can be successfully performed by an adept without it. Instruments : stop speculum (Fig. 72), fixation forceps, bent keratome (Fig. 79), iris forceps (Fig. 81) bent at various angles, according to the position of the iridectomy, iris scissors with elbow bend (Fig. 80), of which some patterns have one or both blades probe pointed, a curette (Fig. 85) for replacing the cut ends of the iris and preventing their incarceration in the angles of the wound. The iridotomy scissors (really shears) are very convenient, especially for downward and inward operations, and for the left hand. A bent or straight broad needle (Fig. 77) may be used instead of the keratome if the iridectomy is to be small.

The conjunctiva is held by the fixation forceps near the cornea at a point opposite to the place selected for puncture. The keratome is to be entered slowly, and steadily pushed on across the anterior chamber till the wound is of the desired size; the knife is then slowly withdrawn, and in its course it may be carefully rotated from side to side so as to make the length of the internal and external wounds equal; or this object may be more certainly gained by using a keratome whose sides become parallel at a certain distance from the point. Two points need attention in making the incision : as soon as the point of the knife is visible in the anterior chamber it must be tilted slightly forwards to avoid wounding the iris and lens, and care must be taken not to tilt it sideways, for by so doing the wound instead of lying parallel with the border of the cornea will lie more or less across it. The incision is made almost as much by pushing the eye against the knife by the fixation forceps, as by pushing the knife against the eye. The fixation forceps are now laid down, or if fixation be still necessary, they are given to an assistant, who is to gently draw the eye in the

FIG. 81.—Bent triangular keratome.

FIG. 80.—Iridectomy scissors.

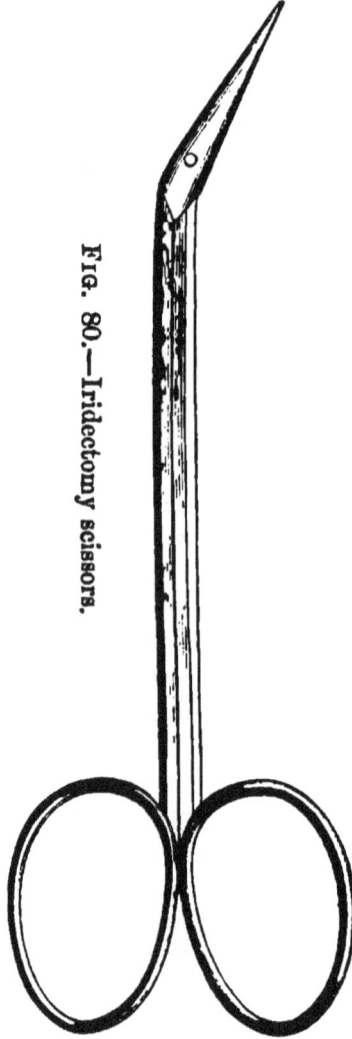

FIG. 79.—Iris forceps.

required direction for facilitating the next step; in so doing he must take care to draw away from the eye, not to push the ends of the forceps against the sclerotic. The iris forceps are now introduced closed into the wound and passed very nearly to the pupillary border of the iris, before being opened and made to grasp the piece to be excised. By seizing the pupillary part of the iris its inner circle is certain to be brought outside the wound, and removed when the forceps are now withdrawn; if the iris be seized in the middle of its breadth a button-hole may be cut out and the pupillary part left standing. Often the iris is carried into the wound by the gush of aqueous as the keratome is withdrawn, and it is then seized without passing the forceps so far into the chamber. The loop of iris having been cut off, either at a single snip, or by cutting first one end and then the other, as in glaucoma (p. 314), the tip of the curette is gently introduced into each angle of the wound to free the iris, should it be entangled; this little precaution is of importance in order to prevent occlusion of the iris in the track of the wound. The speculum is now removed and both eyes bandaged with a pad of cotton wool. We may use either a four-tailed bandage of knitted cotton, or two or three turns of a soft calico or flannel roller.

The anterior chamber is often refilled in twenty-four hours, excepting in cases of glaucoma, when the wound frequently leaks more or less for several days. It is better in all cases to keep the eye bandaged for a week, the wound being but weakly united, and likely to give way from a slight blow or other accident. In cases where the incision lies in the white tissue outside the cornea, some bleeding is apt to occur; when the eye is much congested, the hæmorrhage is considerable, and the blood may run into the anterior chamber either during or after the excision of the iris; the blood can be drawn out by depressing the lip of the wound with the curette, but if the chamber again fills, no prolonged efforts need be made, since it will be absorbed without trouble in a few days. In diseased, especially glaucomatous, eyes secondary hæmorrhage sometimes occurs from the iris several days after the operation, and the absorption of this blood is often slow. When the incision is made with a broad needle it is enlarged by a sawing motion sideways.

Sclerotomy is an operation for dividing the sclerotic near to the margin of the cornea. It is employed by some operators instead of iridectomy in glaucoma, or

FIG. 82.—Diagrammatic section of ciliary region, showing path of wound in iridectomy for glaucoma (*I.*) and in sclerotomy (*S.*)

after a previous iridectomy has failed to stay the progress of this malady. It is performed subconjunctivally, a Graefe's cataract knife (Fig. 83) being entered about a twelfth of an inch (2 *mm.*) from the apparent margin of the cornea, passed in front of the iris and brought out at a corresponding point on the other side so as to include about one third of the circumference; the knife is then made to cut its way through the sclerotic by slight to-and-fro movements, the bridge of conjunctiva not being divided, excepting at the puncture and counter-puncture, and the knife is then withdrawn. The central third or less of the sclerotic flap is usually left undivided. A bluish line is the immediate result, and this often gapes considerably in the next few days. The iris is usually prolapsed more or less into the wound, sometimes so much as to make its removal necessary, when the operation becomes a very peripheral iridectomy. A moderate degree of bulging and separation of the lips of the wound takes place for a week or two, when the scar flattens down, and finally a mere bluish line is left. Sclerotomy is difficult to perform well; if the incision be too long and too far back there is danger of hæmorrhage into the vitreous, and even of puckering and inflammation of the scar and sympathetic ophthalmitis of the other eye; in other cases it may be too short or too far forward, when it is merely equivalent to a partial incision for iridectomy or perhaps even to a paracentesis. In Fig. 82, *I.* shows the line of incision in iridectomy for glaucoma, and *S.* the line in sclerotomy.

G. OPERATIONS FOR CATARACT

Extraction of cataract has been systematically practised for about a hundred years. The operation has passed through many important changes, and many different procedures are still in use. There is also much diversity of practice in regard to anæsthesia, but a large number of the most experienced operators frequently dispense with it. All the operations are difficult to perform well, and much practice is needed to ensure the best prospect of success. Further, the sources of possible failure are numerous, and, since in avoiding one we are very apt to fall into another, it is scarcely likely that any one precise method of operating will ever become universal. At the present time the majority of surgeons adhere with more or less closeness to the operation known as the "modified linear" method of von Graefe.

All operations for extraction agree in the following points: (1.) An incision is made in the cornea or at the junction of cornea and sclerotic, or even slightly in the sclerotic, just large enough to give passage to the crystalline lens without its being broken or altered in shape (*see* Suction Extraction, p. 328). The

21

knife now almost universally employed is
the very narrow, thin, straight knife of von
Graefe (Fig. 83). (2.) The capsule is
freely opened with a small, sharp-pointed
instrument (cystotome or pricker, Fig. 85),
the natural elasticity and brittleness of this

FIG. 83.—Graefe's cataract knife.

membrane causing the rent to gape and
spread. (3.) Removal of the lens through
the rent in the capsule (the latter structure
remaining behind), either by pressure and
manipulation outside the eye, or by the introduction of a traction instrument (scoop
or spoon, Fig. 84) passed behind the lens.

FIG. 84.—Cataract spoon.

Most operators have abandoned the habitual
use of the scoop, reserving it for certain
emergencies and special cases. (4.) Iridectomy is very often performed as the second
stage, not with the primary object of facilitating the exit of the lens, but to lessen
the after risks of iritis; since it has been
found that, where no iridectomy is done,
the portion of iris traversed by the lens
is often so bruised or stretched as to become
the starting-point of severe traumatic iritis.
The iridectomy should be small.

The following are the most important
types of operation at present practised :—
(a.) Graefe's "*modified linear*"* or "*peri-*

* In the "*linear operation*," from which the above
modification has been developed, a much shorter
incision is made with a triangular keratome (Fig. 79) a little

FIG. 85.—Cystotome (upper end), and curette (lower end).

pheral linear " extraction. The incision is slightly beyond the sclero-corneal junction, or partly in the junction and partly beyond it (Fig. 86, 2), and consequently involves the conjunctiva, of which a small flap is made. The incision forms a small arc of a circle much larger than the cornea, and its plane makes a large angle with that of the iris (2, Fig. 87). Iridectomy is performed as the second stage. The incision is made with the long narrow knife of Graefe (Fig. 83), which is so narrow and thin that the direction of its surface carr be easily changed during the progress of the incision. The iridectomy is occasionally made several weeks before the extraction (" preliminary iridectomy "), the parts being allowed to become perfectly quiet in the interval.

A variety of this operation consists in placing the incision rather further down, and at the same time giving it a somewhat sharper curve, so that it forms an arc of a smaller circle than before, but is still not concentric with the cornea (Fig. 86, 3, upper section). The puncture is directed somewhat downwards (as at the right-hand end of the Figure), and its plane, which at the puncture and counter-puncture is almost parallel with the iris, alters to nearly a right angle at the summit of the flap. The track of the wound, if shaded, would appear like the upper section of 3, Fig. 86. Other operators, again, make the whole incision éxactly at the sclero-corneal junction—a modification of the old flap operation known as the short flap; an iridectomy is usually made.

The disadvantages of the peripheral linear extraction are, the frequency of bleeding from the conjunctiva into the anterior chamber, the parts being thus obscured; a greater risk of loss of vitreous, owing to the peripheral position of the wound, and sometimes within the margin of the cornea, and iridectomy is not properly a part of the operation, though it is often advisable. It is applicable only to comparatively soft cataracts, and is now seldom performed.

a difficulty in making the lens present well; a certain risk that the operated eye will set up sympathetic inflammation in its fellow, because a considerable part of the wound lies in the sclerotic, or, short of this, that the operated eye will for long remain irritable from prolapse of the angles of the iris into the corners of the wound and formation of a cystoid scar; lastly, there is a tendency to make the wound rather too short in order to avoid some of these risks, and thus difficulties are introduced in the clean removal of the lens.

(b.) The incision, made with Graefe's knife, has nearly the same curve as in the peripheral linear, but the greater part of the incision lies considerably within the margin of the cornea (*corneal section*) (3, Fig. 86, lower section); and iridectomy is usually

FIG. 86.—Paths of incisions for extraction of cataract. 1, Old flap; 2, peripheral linear; 3 (upper fig.), a variety of the peripheral linear; (lower fig.) corneal section. The wound appears as a narrow slit (2) or a broad track (1), when seen from the front, according to the inclination of its plane. Compare Fig. 87. The dotted circle shows the outline of the lens.

dispensed with. Eserine is sometimes dropped in just before the operation, in order by keeping the pupil contracted for a short time afterwards, to draw the edge of the iris away from the wound, and so prevent prolapse. In Liebreich's operation the section is made downwards, and its plane forms an angle of about 45° with that of the iris (3, Fig. 87). In Lebrun's corneal operation the section is upwards, but is otherwise almost identical with Liebreich's; the plane of the incision, however, varies from about 20° or 30° with the iris at puncture and counter-puncture to a right angle at the centre of the wound. The upper section of 3, Fig. 86, if placed further down in the cornea would nearly

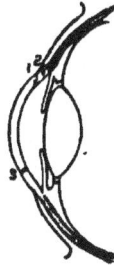

FIG. 87.—The same sections seen in profile, showing the plane of the incision in 1, 2, and the lower section of 3.

represent Lebrun's incision. It remains to be seen whether these corneal operations without iridectomy, which have been adopted by some operators of great experience, are comparatively easy to perform, and usually do not require anæsthesia, will, in the hands of most surgeons, give so high a percentage of useful eyes as Graefe's peripheral linear operation and the short flap operation.

(c.) *Flap extraction.*—The incision is slightly within the visible margin of the cornea, concentric with it, and equal to at least half its circumference (1, Fig. 86), thus forming a large arc of a small circle; and the plane of the incision is parallel with that of the iris (1, Fig. 87). No iridectomy is made. The incision is made with the triangular knife of Beer

FIG. 88.—Beer's cataract knife.

(Fig. 88), in which the blade near its heel is somewhat wider than the height of the flap, and the section completed by simply pushing the knife across the anterior chamber flat with the iris, its back corresponding to the base of the intended flap. The inner length of the wound is less than the outer by the thickness of the obliquely cut cornea at each end (1, Fig. 86).

The after-treatment in flap extraction is troublesome. When everything does well the result is almost perfect, the pupil retaining its natural size, shape, and mobility. The operation is usually done without anæsthetic, and neither speculum nor fixation forceps are needed. The great height of the flap in proportion to its width renders it very liable to gape or even to fall forwards, and this, together with the great extent of the wound in corneal tissue, causes so much interference with nutrition that in a large proportion of cases the cornea sloughs or passes into rapid suppurative inflammation. The iris often prolapses and becomes adherent to the wound, and, even apart from this, severe iritis is a common occurrence. For these reasons the old flap extraction has been almost abandoned in favour of the peripheral linear, corneal section, and short flap operations, which, though giving perhaps a smaller percentage of results that can be called " perfect," yield a much larger average of useful eyes.

Historically, the flap operation was the earliest; then came the linear operation; thirdly, the modified or peripheral linear operation with iridectomy; and lastly, the modern corneal operations and short flap, the aim of which is to gain the substantial advantages both of the old flap and the modified linear methods, without the great risks of the former or the imperfections of the latter.

Some other operations in less general use may be named, e. g. Weber's operation, in which the section is made with a very broad keratome of peculiar pattern; Pagenstecher's in which the lens is removed in its capsule; and Macnamara's, which combines this peculiarity of Pagenstecher's procedure with an incision by a broad keratome and without iridectomy.

After-treatment of extraction by the modified linear, short flap, and corneal operations.—The patient is best in bed for a week. The dressing after the operation consists of a piece of soft linen overlaid by a pad of cotton wool, and kept in place by a four-tailed bandage of knitted cotton ; or the operated eye may be closed by a roller of flannel or soft calico carried once and a half round the head, over the occiput, under the ear and up over the eye. Both eyes are to be bandaged. The room should be kept nearly dark for at least a week, all dressings and examinations being made by the light of a candle. The dressings are removed and the lids gently cleansed with warm water twice a day, their edges being just separated by gently drawing down the lower lid, so as to allow any retained tears to escape; this cleansing is very grateful to the patient. Some surgeons open the lids and look at the eye the day after the operation ; others prefer to leave them closed for several days unless there are signs that the case is doing badly (p. 148). It is a good practice to use one drop of atropine daily after the third day, to prevent adhesions should iritis set in. During the first few hours there will be some soreness and smarting, and at the first dressing a little blood-stained fluid, but after this there should be no material pain nor discomfort, and nothing more than a little mucous discharge, such as old people often have. When first examined (from two to seven days after operation) the eye is always rather congested from having been tied up ; but there should be no chemosis, the wound should be united so as to retain the aqueous, and its edges clear. The pupil is expected to be black unless it is known that portions of lens matter have been left behind. If all is well the bandage *may* be left off during the daytime at the end of the week, a shade being substituted; but it is far safer for the eye to remain tied up for ten days or a fortnight, unless the bandage is very badly borne. The bandage should at first be reapplied at night to prevent accidents from movements during sleep. At the end of a fortnight, if the weather be fine, the patient may begin to go out, with the eyes carefully protected from light and wind, and he may be out of the surgeon's hands in from three to four weeks.

After-operations.—When iritis occurs (p. 149) the pupil becomes more or less occluded by false membrane, which by contracting draws the iris towards the scar, so that the pupil is at once blocked and displaced. In slight cases it is only necessary to tear across the membrane and capsule in the pupil with a fine needle, and to treat the case for a few days as after needle operations for soft cataract. But if the membrane is dense and tough and the iris drawn towards the scar, an artificial pupil must be made, either by iridectomy or by simply cutting the iris and membrane with special scissors, and allowing the tightened tissues to relax (iridotomy). The latter gives excellent results.

Solution and suction operations.—Removal of the lens by *needle operations.*—(1) The pupil is fully dilated by atropine; (2) an anæsthetic is given unless the patient is old enough to control himself well; for the slightest movement is attended by risk; (3) the lids are held open by the fingers, or a stop speculum, and fixation forceps used; (4) a fine cataract needle (Fig. 89) is directed to a point a little within the

FIG. 89.—Cataract needle.

border of the cornea (usually the outer border), and when close to its surface is plunged quickly and rather obliquely through the cornea into the anterior chamber, and its point carried to the centre of the pupil; (5) the point of the needle is dipped back through the lens capsule into the superficial layers

of the lens at the centre of the pupil, and a few
gentle movements made so as to break up the middle
of the front of the lens; (6) the needle is steadily
withdrawn. Special care is to be taken not to wound
nor even touch the iris, either on entering or withdrawing the needle, and not to stir up the lens too
freely.

After-treatment.—The pupil to be kept widely
dilated with atropine (4 grains of the sulphate to
one ounce of distilled water); a drop should be
applied after the operation, and at least three times
a day afterwards, or much oftener if there be threatening of iritis. The eye is to be lightly bandaged
and the patient to remain in bed in a darkened room
for a few days. A little ciliary congestion for two
or three days need cause no uneasiness, but the occurrence of pain and increase of congestion, with
alteration in the colour of the iris (commencing iritis),
are indications for the application of leeches near the
eye, the more frequent use of atropine, and the vigorous use of either ice or an evaporating lotion to
the closed eyelids. Probably iritis would occur much
less often than it does if ice were used from the beginning in all cases.*

If the cataract were complete, no marked change
will be seen for some weeks; if partial (*e.g.* lamellar),
the neighbourhood of the needle wound will become
opaque in two or three days. In from six to eight
weeks the lens will have become notably smaller
(flattened or hollowed on the front surface). If the
eye be pefectly quiet, but not unless, the operation
may now be repeated in exactly the same way, and
with the same after-treatment and precautions, but
the needle may be used more freely. The bulk of the
lens will generally disappear after the second opera-

* I have to thank Mr Gunn, the able house surgeon of the
Moorfields Hospital, for this suggestion, which has, I believe,
been carried out by him for some time past with the best
results.

tion, but the needle often needs to be used a third or a fourth time for the distintegration of small residual pieces, or in order to tear the capsule if it has not retracted enough to leave a clear central pupil. A small whitish dot often remains in the cornea at the seat of the needle puncture.

Removal of the lens by suction.—This operation is applicable to soft cataracts, but unless very skilfully performed, is attended by serious danger. The eye is thoroughly atropised, and an oblique opening made in the cornea with a broad, cutting needle (Fig. 77), between its centre and margin, and the lens capsule freely lacerated. The needle being withdrawn, the nose of the syringe is passed through the wound and gently dipped into the lacerated lens substance. Very gentle suction is now used, and the semifluid lens-matter drawn gradually into the syringe. The instrument is not to be passed behind the iris in search of fragments. When the operation is successful, the whole lens is removed at one sitting. The after-treatment is the same as for needle operations. Two forms of syringe are in use; in the one originally introduced by Mr Teale, the suction is made by the mouth applied to a piece of flexible india-rubber tubing; in the other, by Mr Bowman, the suction is obtained by a sliding piston worked by the thumb moving along the length of the syringe. It is often better, and in lamellar cataract necessary, to break up the lens freely with a fine needle a few days before using the syringe, and thus allow it to be thoroughly macerated and softened in the aqueous. When this course is taken the patient must be carefully watched, and kept in a darkened room, and atropine be used freely between the needle operation and the suction; and the surgeon must be prepared to interfere before the day appointed for the suction should inflammatory symptoms be set up by the rapid swelling of the lens.

PART III

CHAPTER XXII

In stating very shortly the most important facts
bearing on the connection between diseases of the eye
and of other parts of the body, it is convenient to
distinguish the following groups of cases :—(A) in
which the eye changes occur as part of a general
disease; (B) in which the ocular disease is symptomatic of some local malady at a distance; (C)
cases where the eye shares in a local process, affecting
the neighbouring parts.

(For the clinical details of the various eye diseases
referred to in this chapter, *see* Part II.)

A. General diseases, in which the eye is liable to
suffer.

Syphilis is, directly or indirectly, the cause of a
large proportion of the more serious diseases of the
eye.

Acquired syphilis.—Primary stage. Hard chancres
are occasionally seen on the eyelid. I have once seen
one far back on the conjunctiva.

Secondary stage (sore throat, shedding of hair,
eruption, and condylomata). — *Iritis* is common
between two and eight or nine months, and does not
occur later than about eighteen months, after the
contagion; in a large proportion (more than half the

cases) both eyes suffer; there is a marked tendency
to exudation of lymph (plastic iritis), shown by keratitis punctata, haze of cornea, and less commonly by
lymph-nodules on the iris. In some cases there are
symptoms of severe cyclitis with but little iritis ; but
the cyclitis of acquired syphilis does not give rise to
ciliary staphyloma. Syphilitic iritis, though sometimes protracted, rarely relapses after complete sub-
sidence. *Choroiditis* and *retinitis* generally set in
rather later, from six months to about two years
after the chancre. The two conditions are most often
seen together, but either choroiditis or retinitis may
occur singly ; and in either event the vitreous generally becomes inflamed. These conditions are essentially chronic, the retinitis being often, and the
choroiditis sometimes, liable to repeated exacerbations
or recurrences; whilst in some cases the secondary
atrophic changes progress slowly for years, almost
to blindness, often with pigmentation of the retina.
Syphilitic choroiditis and retinitis usually affect both
eyes, but often in an unequal degree.* In a few
cases detachment of the retina and secondary cataract
occur in secondary syphilis. *Keratitis*, indistinguishable from that of inherited syphilis, is amongst the
rarest events in the acquired disease, and when seen
always forms part of the secondary stage.

Tertiary stage.—Ulceration of the skin and conjunctiva of the lids, gummatous infiltration of the
lids, and nodes in the orbit (whether cellular or
periosteal) occur but rarely. *Oculo-motor paralysis*
is one of the commonest ocular results of syphilis.
It may depend upon gumma (syphilitic neuroma) of
the affected nerve in the orbit or in the skull, or on
gummatous inflammation of the dura mater at the

* Choroiditis sometimes occurs at a later stage, in only
one eye, and without retinitis, when it deserves to be classed as
a tertiary symptom. But these cases are, I believe, much less
common than the symmetrical choroiditis (or choroido-retinitis)
of secondary syphilis.

base of the skull, matting the nerves together, or on disease of nerve centres, causing ophthalmoplegia externa. The gummatous nerve lesions seldom occur very late in tertiary syphilis.

Diseases of the optic nerve in relation to syphilis.—The retinitis of the secondary stage affects the disc, and when atrophy of the retina and choroid occur the disc becomes wasted in proportion; in rare cases the retinitis of secondary syphilis is replaced by well-marked papillitis of local origin. Such cases must not be confused with others, equally rare, in which double papillitis, passing into atrophy, occurs with all the symptoms of severe meningitis in secondary syphilis. Tertiary syphilitic disease, anywhere within the cranium, commonly causes papillitis, in the same way as do other coarse intra-cranial lesions; but gummatous inflammation of the trunk of the optic nerve, or of the chiasma, may also be the cause of descending neuritis. Primary progressive atrophy of the discs occurs in association with syphilitic locomotor ataxy and ophthalmoplegia externa; probably in a few instances it occurs alone, or for a time precedes the other changes, as is known to be the case in the ataxic diseases not caused by syphilis.

Inherited syphilis.—Secondary and tertiary stages. —*Iritis* corresponding to that in the acquired disease is seen in a small number of cases, and occurs between two and about fifteen months of age. It often gives rise to much exudation, leading to occlusion of the pupil, and is frequently accompanied by deeper changes. It is very often symmetrical, and is much commoner in girls than boys. *Choroiditis* and *retinitis*, of precisely the same forms as in acquired syphilis, occur at the corresponding period of the disease, *i. e.* between six months and about three years of age; and they show as much (some observers think more) tendency to the degenerative and atrophic results already described. *Keratitis* is the commonest eye disease caused by inherited syphilis. It seldom comes on

before the tertiary period, the great majority of its
subjects being between seven and fifteen years old,
but it is sometimes seen as early as two or three
years,* and may be deferred till about twenty-five.
The disease always lasts for months, is frequently
complicated with iritis and cyclitis, and, though tend-
ing to recovery, shows a considerable liability to
relapse.   It almost always attacks both eyes, though
sometimes at an interval of many months.   When
occurring unusually early its course is generally
short and mild.   The *oculo-motor palsies* occur but
rarely in inherited syphilis; a case in a young child
has been recorded, with post-mortem examination, by
Dr Barlow.

Smallpox causes inflammation and ulceration of
the cornea, leading, in the worst cases, to total destruction, but in a large number to nothing worse
than a chronic vascular ulcer.   The corneal disease
comes on some days after the eruption (tenth to
fourteenth day after its commencement), and after
the onset of the secondary fever.   Iritis, uncomplicated and showing nothing characteristic of its cause,
sometimes occurs some weeks after an attack of
smallpox.   Only in very rare cases do variolous pustules form on the eye, and even then they are always
on the conjunctiva, not on the cornea.

Scarlet fever, typhus, and some other exanthemata
may be followed by rapid and complete loss of sight,
lasting a day or two, showing no ophthalmoscopic
changes, and ending in recovery.   Such attacks are
believed to be uræmic, or at any rate dependent on
some toxic condition of the blood.   A peculiarity of
these cases is the preservation of the action of the
pupils to light.   Scarlet fever is sometimes accompanied or followed by diphtheritic ophthalmia.

Diphtheria.—By far the commonest result is paralysis (oftenest incomplete) of the ciliary muscles

* I have lately seen a case of severe double diffuse keratitis
in a very syphilitic infant of about *three months* old.

(*cycloplegia*) ; the irides are not affected excepting
in very severe cases, when the pupils may be rather
large and sluggish. The symptoms come on from
four to six weeks after the commencement of the
illness, and generally last about a month, complete
recovery ensuing. Diphtheritic cycloplegia is most
commonly, but not invariably, accompanied by paralysis of the soft palate. In most of the cases seen
by ophthalmic surgeons, the attack of diphtheria has
been mild, sometimes extremely so, the case often
being described as "ulcerated throat;" but inquiry
often yields a history of other and severer cases in
the family ; and of general depression and weakness
in the patient, out of proportion to the throat symptoms. We find that most of the patients who apply
with diphtheritic cycloplegia are hypermetropic,
doubtless because those with normal (and, *à fortiori*,
with myopic) refraction are much less troubled by
paresis of accommodation, and often do not find it
necessary to seek advice. Concomitant convergent
squint is sometimes developed in hypermetropic
children during the diphtheritic paresis, owing to
the increased efforts at accommodation, and this
has probably led to the statement that paralysis of
the external rectus is caused by diphtheria; paralysis
of the external muscles of the eye, if it ever occur,
is certainly very rare.

Diphtheritic and membranous ophthalmia are sometimes caused by direct infection of the conjunctiva by
diphtheritic material from the throat of another
person ; while in other cases the diphtheritic process
creeps up the nasal duct from the nose, and thus
reaches the lining membrane of the eyelids. But
in the majority of cases of "diphtheritic" and
"membranous" ophthalmia the disease is a local
one, in which the inflammation takes on this special
form. No doubt there is often something peculiar in
the patient's health, or in the state of his eye-tissues,
which gives a proclivity to this particular patho-

logical process; these cases are, for instance, seen with particular frequency after measles, and less commonly during or after scarlet fever, and are more likely to occur in children than adults; the existence of old granular disease of the conjunctiva also gives a strong tendency to a diphtheritic type of inflammation, and the same tendency is seen sometimes in a well-marked degree in ophthalmia neonatorum and in gonorrhœal ophthalmia. As there seems but seldom any reason to look upon these forms of ophthalmia as the local manifestations of a specific blood disease, the term "diphtheria of the conjunctiva" should, I think, seldom be used. Many of the best-marked membranous cases occur sporadically, and without any history suggestive of infection from throat diphtheria or from any other exanthem.

Measles is a prolific source of ophthalmia tarsi in all its forms, and of corneal ulcers, particularly of the phlyctenular forms. It also gives rise to a troublesome muco-purulent ophthalmia, and under bad hygienic conditions this appears liable to become aggravated by cultivation and transmission, into destructive disease of purulent, membranous, or diphtheritic type.

Chicken-pox is sometimes followed by a transient attack of mild conjunctivitis.

Whooping-cough often, like measles, leaves a proneness to corneal ulcers. In a few rare cases the condition known as *ischæmia retinæ* (sudden temporary arterial bloodlessness) has occurred.

Malarial fevers, especially the severe forms met with in hot countries, are sometimes the cause of retinal hæmorrhages (often large and periarterial), and even of considerable neuro-retinitis; when there is much pigment in the blood, the swollen disc may have a peculiar grey colour. When renal albuminuria is caused by malarial disease, albuminuric retinitis may occur.

Relapsing fever is sometimes followed during convalescence by choroiditis (evidenced by opacities in the vitreous), with or without iritis; severe inflammatory symptoms are sometimes present, but recovery takes place. The eye changes seem to be much commoner in some epidemics than in others.

Epidemic cerebro-spinal meningitis also, in a few cases, gives rise to acute choroiditis, with pain, chemosis, and great tendency to rapid exudation of lymph into the vitreous and anterior chambers, and often leading to disorganisation of the eye and blindness.* It is believed that the inflammation may either extend to the eye along the optic nerve or may occur independently in the brain and the eye. Deafness from disease of the internal ear is still commoner than the eye disease.

Purpura has been observed in a few cases to be accompanied by retinal or subretinal hæmorrhages; they are sometimes perivascular and linear, and in other cases form large blotches. They have also been found in *Scurvy*.

Lead poisoning is an occasional cause of optic neuro-retinitis leading to atrophy, of primary atrophy, and of toxic amblyopia (rapid and usually transient). The two former are the most common; the atrophy, whether primary or consecutive to papillitis, is generally accompanied by very marked shrinking of retinal arteries, and great defect of sight or complete blindness; it is generally symmetrical, but one eye may precede the other. Other symptoms of lead poisoning are present in nearly all cases, usually chronic, but occasionally acute. Care must be taken not to confuse albuminuric retinitis from kidney disease induced by lead, with the changes here alluded to and which are due in some more direct manner to the influence of the metal.

* Possibly some of the rare cases in which similar eye conditions are seen without apparent cause may be the accompaniments of slight and unrecognised meningitis (*see* p. 245).

22

The deposition of lead upon corneal ulcers has been referred to at p. 96.

Alcohol.—There is a widely-spread belief amongst ophthalmic surgeons that alcohol, especially in the form of distilled spirit, causes a particular form of symmetrical amblyopia (the so-called *amblyopia potatorum*). The difficulty of arriving at the truth is immensely increased by the fact that most drinkers are also smokers, and that tobacco, whether smoked or chewed, is allowed by all observers to be also one of the causes (or, as some hold, the sole cause) of a similar disease. For the most part but little trouble is taken to distinguish between the two conditions, or even to treat of them as clinically distinct; Dr Hirschberg, however, has attempted to show that the defect of the visual field is somewhat different in the two, and that this difference may be utilised in the differential diagnosis. The question of whether alcohol directly causes disease of the optic nerves will not be settled until observers are much more careful than they have hitherto been to record as typical cases of alcoholic amblyopia only those in which the patient has never used tobacco in any shape, nor in the smallest quantity. Magnan thinks alcoholic amblyopia less common than some have supposed.*

Failure of one eye only, leading to atrophy of the disc, is described as sometimes the result of chronic alcoholism.

Tobacco.—Whatever may be the truth (and it is confessedly difficult to arrive at) as to the direct influence of alcohol, and of the various substances often combined with it, there is no doubt whatever that tobacco, whether smoked or chewed, does act directly on the optic nerves, and in such a manner as to give rise to definite, and usually very characteristic symptoms. The amblyopia seldom comes on until tobacco has been used for many years. The daily

* Magnan 'On Alcoholism,' Greenfield's translation, p. 42.

quantity needed to cause symptoms is, *cæteris paribus*, a matter of idiosyncrasy, and very small doses will produce the disease in men who in other respects are unable to tolerate large quantities of the drug. Predisposing causes exert a very important influence : amongst these are to be especially noted increasing age; nervous exhaustion from overwork, anxiety, or loss of sleep; chronic dyspepsia, whether from drinking or other causes; and probably sexual excesses and exposure to tropical heat (or light ?). A large proportion of the patients drink to excess, and thus make themselves more susceptible to tobacco, both by injuring the nervous system and the stomach. But some remarkable cases are seen in men who have for long been total abstainers, in others who have lately become abstainers without lessening their tobacco, and in yet others who are strictly moderate drinkers and in whom increasing age is the only recognisable predisposing cause. The strong tobaccos produce the disease more readily than the weaker sorts, and chewing is more dangerous than smoking.

Quinine, taken in large doses, at short intervals, has in a few cases caused serious failure of sight, with central defect of field (or, according to some, peripheral contraction). Recovery is sometimes rapid, but may be delayed for months. The amblyopia is not always equally symmetrical. There are no opthalmoscopic changes.

Kidney disease.—The common and well-known retino-neuritis, associated with renal albuminuria, and of which several clinical types are found, has been already described. It need only be added that the disease is commonest with chronic granular kidneys and in the kidney disease of pregnancy, but that it is also seen in the chronic forms following acute nephritis and in lardaceous disease; and that it is but rarely seen in children. The disease may relapse. Detachment of the retina is an occasional result in extreme cases.*   The prognosis as regards vision is best in the cases depending on albuminuria of pregnancy. The retinitis is intimately associated with the albuminuria, though the nature of the connexion is obscure; it is not caused by the cardiac hypertrophy which is so often present. The failure of sight caused by albuminuric retinitis has often led to the correct diagnosis of cases which had been treated for dyspepsia, headache, or "biliousness."

In some cases of chronic renal disease with high arterial tension the retinal arteries have been observed by Dr Gowers to be markedly contracted without any retinitis.

Diabetes sometimes causes cataract. In young or middle-aged patients the cataract usually forms quickly, and is, like other cataracts at the same ages, soft. As it is always symmetrical, the rapid formation of double complete cataract, at a comparatively early age, should always lead to the suspicion of diabetes. In old persons the progress of diabetic cataract is much slower, and often shows no distinctive characters. The relation of the lenticular opacity to the diabetes has not been satisfactorily explained : the presence of sugar in the lens ; the action of sugar or its derivatives dissolved in the aqueous and vitreous ; the abstraction of water from the lens owing to the increased density of the blood ; and, lastly, degeneration of the lens from the general cachexia attending the disease, have all been offered in explanation.

In a few cases retinitis occurs attended by great œdema and copious (probably capillary) hæmorrhages into the retina and vitreous.

Leucocythæmia is often accompanied by retinal hæmorrhages, less commonly by whitish spots bordered by blood, and consisting of white corpuscles ; these spots may be thick enough to project forwards. Occasionally there is general haziness of the retina.

---

* I have seen this result in three patients.

These changes are usually symmetrical. In severe cases the whole fundus is remarkably pale, whether there be other changes or not.*

Progressive pernicious anæmia is marked by a strong tendency to retinal hæmorrhages; these are usually grouped chiefly near the disc, and are striated (Gowers). White patches are also common, and occasionally well-marked neuritis occurs. I have seen hæmorrhages of different dates, and in one case, shown to me by Dr. Sharkey, there had evidently been a large extravasation from the choroid at an earlier period. The disc and fundus participate in the general pallor.

Heart disease is variously related to changes in the eyes and alterations of sight. Aortic incompetence often produces visible pulsation of the retinal arteries. This pulsation differs from that seen in glaucoma by extending, in many cases, far beyond the disc, and in not being so marked as to cause complete emptying of the larger vessels during the diastole. In glaucoma, the pulsation is confined to the disc, and often amounts to complete emptying of the arteries. The difference is explained by the different mode of production in the two cases; in the one the incomplete closure of the aortic orifice, by lowering the pressure in the whole blood-column during the diastole, allows a reflux of blood; in the other increased resistance to the entrance of arterial blood, due to heightened intraocular tension, acts chiefly on the comparatively yielding tissues of the optic disc. Valvular disease of the heart is generally present in the cases of sudden lasting blindness of one eye, clinically diagnosed as embolism of the arteria centralis retina, but in some of which thrombosis of the same artery or its companion vein, or blocking of

* For a full account of the changes see Gowers' 'Medical Ophthalmoscopy.' Dr. Sharkey has lately shown me a case with diffuse retinitis, very numerous punctiform hæmorrhages, chiefly peripheral, and dilatation with extreme tortuosity of the veins.

the internal carotid* and ophthalmic arteries, have
been found *post-mortem*. Brief temporary failure or
loss of sight is not uncommon in the vascular disturbances to which the subjects of valvular heart disease
are subject.   It is probable that high arterial tension
predisposes to intraocular hæmorrhage in cases where
the small vessels are unsound, and that the frequent
association of retinal hæmorrhage with cardiac disease is thus explained.

Acute generalised tuberculosis is in a few cases accompanied by the growth of miliary tubercles in the
choroid; they are said not to occur in tubercular
meningitis without general tuberculosis.   Chronic
large growths of confluent tubercles are occasionally
seen in the eye, and may simulate malignant tumours.   There is reason to suspect that choroidal
tubercles sometimes form in cases of tubercular meningitis which recover, and that certain cases of localised choroiditis not accompanied by serious general
symptoms may be of tubercular character.

Rheumatism.—In acute rheumatism Dr Barlow
informs me that he has more than once seen well-
marked congestion of the eyes and photophobia; I
believe, however, that neither iritis nor other inflammatory changes ever occur.   But the subjects of
chronic articular rheumatism are very subject to relapsing iritis.   Some of these patients give a history
of acute rheumatism as the starting-point of their
troubles, others of a prolonged attack of subacute
character lasting for many months, whilst in a third
group the articular symptoms have never been severe.
In yet another series a liability to fascial or muscular
rheumatism, or to recurrent neuralgia from exposure
to cold or damp, are the only " rheumatic" symptoms
of which a history is given; in some of these the
neuralgia is probably gouty.   It is to be remembered that the eye is now and then the first part to
be attacked by an inflammation, which later events

* Gower's ' Medical Ophthalmoscopy, p. 27, foot-note.

show to be clearly related to rheumatism or to gout.

Gonorrhœal rheumatism is not unfrequently the starting-point of relapsing iritis and chronic relapsing rheumatism. Rheumatic iritis occurring for the first time in the primary attack of gonorrhœal rheumatism is, in my experience, more often symmetrical than other forms of arthritic iritis, or than the later attacks of iritis in the same patient; a fact which sometimes makes the distinction between rheumatic and syphilitic iritis difficult.

It is believed that rheumatism is the cause of some cases of non-suppurating orbital cellulitis, and of relapsing episcleritis. Rheumatism is also believed to cause some of the ocular paralyses.

Gout.—Gouty persons are not very unfrequently the subjects of recurrent iritis indistinguishable from that which occurs in rheumatism. Rheumatism and gout seem sometimes so mixed that it is not always possible to assign to each its right share in the causation of iritis; but that the subjects of true " chalk gout" are liable to relapsing iritis is undoubted. There is, on the whole, more tendency to insidious forms of iritis in gout than in rheumatism. It is also generally believed, and I think with reason, that the subjects of gout, or persons whose near relatives suffer from it, are particularly subject to glaucoma; acute glaucoma was indeed the " arthritic ophthalmia," of earlier authors. Hæmorrhagic retinitis is also commoner in gouty persons than in others; it is usually unsymmetrical, and to be distinguished from double albuminuric retinitis, which may occur in the same subjects.

It has also been observed that the children or descendants of gouty persons, without being themselves subject to gout, are sometimes attacked in early adult life by an insidious form of irido-cyclitis often leading to secondary glaucoma and serious damage to sight;* both eyes are attacked sooner or later. The cases in this group probably seem rarer than they are, from the impossibility in many instances of getting a full family history. .

Several different clinical types may be recognised in the large group of maladies referred to in this section under the name of "iritis." Besides the purely iritic cases, we may distinguish some as cyclitic, sometimes with increase, in other cases with decrease of tension; in another group the sclerotic and conjunctiva are chiefly affected (true "rheumatic ophthalmia" without iritis); a fourth group, in which the pain is disproportionately severe, may be spoken of as neuralgic. In a large majority, however, the iris is the headquarters of the morbid action. All arthritic eye diseases (excepting hæmorrhagic retinitis) are marked by a strong tendency to relapse; they usually attack only one eye at a time, though both suffer sooner or later; and they are all much influenced by conditions of weather, being commonest in spring and autumn.

The strumous condition is a fruitful source of superficial eye diseases, which are for the most part tedious and relapsing, are often accompanied by severe irritative symptoms, but, as a rule, do not lead to serious damage. The best types are—(1) the different varieties of ophthalmia tarsi; (2) all forms of phlyctenular ophthalmia ("pustular" or "herpetic" diseases of the cornea and conjunctiva); (3) many superficial relapsing ulcers of cornea in children and adolescents, though not distinctly phlyctenular in origin, are certainly strumous; (4) many of the less common, but very serious, varieties of cyclo-keratitis in adults occur in connection with lowered health, susceptibility to cold, and sluggish but irritable circulation, if not with decidedly scrofulous manifestations; (5) lupus is, of course, a strumous disease, whether attacking the parts around the eye or other parts.

* Hutchinson, the 'Lancet,' Jan., 1873.

Entozoa sometimes come to rest and develop in the eye or orbit. The commonest intraocular parasite is the *cysticercus cellulosæ;* it is excessively rare in this country, but commoner on the Continent. The cysticercus may be found either beneath the retina, in the vitreous, or upon the iris, and may sometimes be recognised in each of these positions by its movements. The parasite has been successfully extracted from the vitreous; when situated on the iris its removal involves an iridectomy. Sometimes it develops under the conjunctiva, where it may set up suppurative inflammation. The *ecchinococcus* hydatid with multiple cysts may develop to a large size in the orbit, and cause much displacement of the eyeball.

B. Eye disease, or eye symptoms, indicative of local disease at a distance.

Megrim is well known to be sometimes accompanied or even solely manifested by temporary disorder of sight. This generally takes the form of a flickering cloud (" flittering scotoma " of German authors) with serrated borders, which, beginning near the centre of the field, spreads eccentrically so as to produce a large defect in the field, a sort of hemiopia; the borders of the cloud may be brilliantly coloured. It affects both eyes, and is visible when the lids are closed. The attack lasts only a short time, and perfect sight returns. In many patients this amblyopia is the precursor of a severe sick headache, but in others it constitutes the whole attack; it never follows the headache. Less definite and characteristic symptoms (dimness, cloudiness, or muscæ) are complained of by some patients. Eye symptoms from megrim are said to be commoner in advanced life.*

Neuralgia of the fifth nerve, especially of its first division, in a few cases precedes or accompanies

* Clifford Allbutt, 'British and Foreign Medico-Chirurgical Review,' 1874.

blindness of the corresponding eye with neuritis or
atrophy of the disc. A liability to neuralgia of the
face and head is not unfrequently observed in persons
who subsequently have glaucoma (compare p. 342
and 344).

Intense neuralgic pain in the face or head sometimes causes dimness of sight of the same eye, whilst
the pain lasts.

I have not hitherto seen any evidence supporting
the belief that injury to branches of the fifth nerve
can cause amaurosis.

Sympathetic ophthalmitis should be mentioned
here as the best known instance in which inflammation of the eyeball is caused by local disease of a
distant part.

In cases of *hysterical hemianæsthesia* there is
temporary defect or blindness of the eye on the same
side, with colour blindness. Such cases rarely come
to the notice of the ophthalmic surgeon.

Cases of partial symmetrical paralysis of the third
nerve are sometimes seen in women without recognisable organic cause. Usually only the upper lids
are affected (symmetrical incomplete ptosis), but
other muscles may be involved. Its subjects are
usually women of susceptible nervous temperament,
and there is often a history of long-continued overwork or anxiety. Many of the cases of *accommodative asthenopia*, without apparent cause, occur in
women with disordered menstruation.

Diseases of the central nervous system may be shown
in the eye either at the optic disc (papillitis and atrophy), or in the muscles (strabismus and diplopia).

The diseases which most often cause *papillitis* are
intracranial tumours, syphilitic growths, and meningitis. Abscess of the brain and softening from embolism and thrombosis are less common causes of
papillitis, and cerebral hæmorrhage scarcely ever
causes it.

In a very large proportion* of all the cases of *cerebral tumour* (excluding syphilitic growths) neuritis occurs at some period. The severity and duration of the neuritis vary much, and probably depend in many cases on the rate of progress as well as on the character of the morbid growth. It not uncommonly sets in at no long interval before death, whilst in some cases it is very chronic. There is nothing in the characters or course of the neuritis to help us in the localisation of intracranial tumour; and excepting that a very high degree of neuritis, with signs of great obstruction to the retinal circulation, generally indicates cerebral tumour, the pathological character of the intracranial disease, whether tumour, meningitis, or syphilitic disease, is not much elucidated by the mere occurrence of papillitis. Tumours also sometimes cause simple optic atrophy by pressing upon or invading some part of the optic fibres.

Intracranial syphilitic disease is a common cause of papillitis, the disease being either a gummatous growth in the brain, or a growth or thickening beginning in the dura mater, or basilar meningitis. The prognosis is much better than in cerebral tumours if vigorous treatment be adopted early, and in all cases of papillitis, where intracranial disease is diagnosed and syphilis even remotely possible, full doses of iodide of potassium should be promptly given.

Meningitis often causes papillitis, but in this respect much depends on its position and duration. Meningitis limited to the convexity, whatever its cause, is seldom accompanied by ophthalmoscopic changes; on the other hand, basilar meningitis very often causes neuritis. The neuritis in basilar meningitis is probably proportionate to the duration and intensity of the intracranial mischief, being comparatively slight in acute and rapidly fatal cases, whether they be tubercular or not. In tubercular cases the dis-

* Dr Gowers thinks in four fifths.

ease seems especially related to the occurrence of inflammatory changes about the chiasma (Gowers) ; and the neuritis in cases of cerebral tumour also seems sometimes to be caused by secondary meningitis set up by the growth. When patients recover from meningitis the neuritis may pass into atrophy and cause amaurosis; such cases are commonest in children, and form an important group, well known to ophthalmic surgeons; it is probable that some of them may be instances of recovery from tubercular meningitis.

In rare cases papillitis occurs with severe head symptoms, ending in death, but without macroscopic changes in the brain or membranes. Microscopical changes in the brain substance, justifying the term cerebritis, have been found in one such case by Dr Sutton, and in another by Dr Stephen Mackenzie. It is to be remembered that optic neuritis may be caused by various altered conditions of the blood, and that it is occasionally seen without any evidence either of central nervous disease or of a morbid state of the blood.

Hydrocephalus rarely causes papillitis, but often at a late stage causes atrophy of the optic nerves from the pressure of the distended third ventricle on the chiasma. Dr Barlow informs me that in several cases he has observed a very gross form of choroiditis ending in immense patches of atrophy, and I have also recorded one such case.

There appears to be an intimate relation between the occurrence of *convulsions* and the formation of lamellar cataract, this form of cataract being scarcely ever seen except in those who have had fits in infancy. A very striking deformity of the teeth is also nearly always present, depending upon an abruptly limited deficiency or absence of the enamel on the upper part of the tooth. The teeth affected are the first molars, incisors, and canines, of the permanent set. The dental changes are quite different from those which are pathognomonic of inherited syphilis, although mixed forms are sometimes seen. The relation between the convulsions, the cataract, and the defective dental enamel has not been satisfactorily explained. Mr Hutchinson has collected many facts in favour of the belief that the dental defect is due to stomatitis interfering with the calcification of the enamel before the eruption of the teeth, and that mercury is the commonest cause of this stomatitis. On this hypothesis the coincidence of the dental defect and the cataract is due to mercury having been usually prescribed for the infantile convulsions from which these cataractous children suffer. There also seems, however, much probability in the supposition that the defect of the crystalline lens and of the enamel, both of them epithelial structures, may be caused by some common influence; although the facts that the peculiar teeth are often seen without the cataract, and the cataract occasionally seen with perfect teeth, appear to weaken this view.

The diseases most commonly causing *atrophy not preceded by papillitis* are the chronic progressive diseases of the spinal cord, especially locomotor ataxy. The atrophy in these cases is slowly progressive, double, though seldom beginning at the same time in both eyes, and it always ends in blindness, although sometimes not until after many years. Similar atrophy sometimes occurs in the early stages of general paralysis of the insane, but in the only two cases of this kind that I have seen ataxic symptoms were also present. It is also, but much more rarely, seen in lateral and in insular sclerosis. In the last-named disease, amblyopia without ophthalmoscopic changes is occasionally seen.

Motor disorders of the eyes.—Some of the commoner causes of ocular palsy have been already given. It may be mentioned here that basilar meningitis

often causes paralysis of one or more of the ocular
nerves with squinting (and double vision if the
patient be conscious), and, further, that the palsy in
such cases often varies, or appears to vary, from day
to day. Locomotor ataxy and general paralysis of
the insane are sometimes preceded. by temporary
paralysis of one or more of the eye muscles, causing
diplopia, and there may for some years be nothing
else to attract the patient's attention ; the same diseases may also be preceded by paralysis of the iris
and ciliary muscle. Paralysis of one third nerve
coming on at the same time as hemiplegia of the
opposite side may point to disease of the crus cerebri
on the side of the palsied third. Loss of one of the
associated or conjugate movements of the eyes (e. g.
loss of power to look to the right or left) is, in most
cases, only one among much graver symptoms in
recent hemiplegia ; but a few such cases have been
seen in which, with the exception of headache and
vomiting, there were no serious symptoms (Priestley
Smith).

Insular (disseminated) sclerosis is often accompanied by nystagmus, characterised by the irregularity, both in amplitude and rapidity, of the
movements.

C. Cases in which the eye shares in a local process
affecting the neighbouring parts.

In *herpes zoster* of the first division of the fifth
nerve the eye participates. When only the supraorbital or supra-trochlear branches are attacked,
the eyeball escapes, or is only superficially congested and the lid swollen. But if the eruption
occurs on the parts supplied by the nasal or infra-
trochlear branches (*i. e.* if the spots extend down
to the tip of the nose), there is commonly inflammation of the proper tissues of the eyeball
(ulceration of cornea and iritis) ; for the sensitive nerves of the cornea, iris, and choroid are

derived, through the long root of the ophthalmic ganglion, from the nasal branch. The pain and swelling of the herpetic region are often so great that the attack gets the name of "erysipelas." In rare cases paralysis of the third nerve accompanies the herpes.

In *paralysis of the first division of the fifth* the cornea and conjunctiva are anæsthetic; the cornea may be touched or rubbed without the patient feeling at all. In many cases ulceration of the cornea, usually uncontrollable and destructive in character, takes place. It is doubtful whether this is due directly to paralysis of trophic fibres running in the trunk of the fifth, or indirectly to the anæsthesia. The anæsthesia operates first by allowing injuries and irritations to occur unperceived, and, secondly, by removing the reflex effect of the sensitive nerves on the calibre of the blood-vessels, and thus permitting inflammation to go on uncontrolled.

In *paralysis of the facial nerve* the eyelids cannot be shut, and the cornea remains more or less exposed. When a strong effort is made to close the lids the eyeball rolls upwards beneath the upper lid. Epiphora is a common result of facial palsy. Severe ulceration of the cornea may occur from the exposure.

Paralysis of the cervical sympathetic causes some narrowing of the palpebral fissure, apparent recession of the eye into the orbit, and non-dilatability (rather than contraction) of the pupil. No changes are observed in the calibre of the blood-vessels of the eye. The pupil is said to be less contracted after division of the sympathetic trunk than when the trunk of the fifth (and with it the oculo-sympathetic fibres) is cut, and knowledge of this may be now and then useful in diagnosis.

In *exophthalmic goitre* the eyeballs are too prominent, and the protrusion, though not always quite equal, is almost invariably bilateral. It is often ap-

parently increased in slight cases by an involuntary
and excessive retraction of the upper lids, especially
when the patient looks down. In severe cases the
proptosis may be so great as to prevent full closure
of the lids, and in these, ulceration of the cornea is
to be feared. In such cases it is beneficial to shorten
the palpebral fissure by uniting the borders of the
lids at the outer canthus, or to divide the tendon of
the levator palpebræ. No changes are present in the
fundus, excepting sometimes dilatation of arteries
and spontaneous arterial pulsation.

Erysipelas of the face sometimes invades the deep
tissues of the orbit and causes blindness by affecting
the optic nerve and retina. On recovering from the
erysipelas in such a case the eye is found to be blind
and the ophthalmoscope shows either simple atrophy
of the disc, or signs of past retinitis also. Other
forms of orbital cellulitis may lead to the same
result.

FORMULÆ, ETC.

NITRATE OF SILVER.

Mitigated Solid Nitrate of Silver:
> Nitrate of Silver 1,
> Nitrate of Potash 2.

Fused together and run into moulds to form short, pointed sticks.

Used for granular lids and purulent ophthalmia.

The strength above given is known as No. 1, and is that which I generally use; three weaker forms are made, known as Nos. 2, 3, and 4, containing respectively 3, 3½, and 4 parts of Nitrate of Potash to 1 of Nitrate of Silver.

Pure Nitrate of Silver is never to be used to the conjunctiva.

Solutions of Nitrate of Silver:
> (1) Nitrate of Silver gr. x,
> Distilled Water ℥j.

Used by the surgeon for purulent ophthalmia, recent granular lids, and some ulcers of cornea.

> (2) Nitrate of Silver gr. j or ij,
> Distilled Water ℥j.

Used by the patient in various forms of acute ophthalmia; only a few drops to be used at a time, and not more than three times a day.

All Solutions of Nitrate of Silver should be kept either in a deep blue bottle, or in a dark place.

SULPHATE OF COPPER.

A crystal of *Pure Sulphate of Copper*, smoothly pointed, may be used for touching granular lids of old standing.

Lapis Divinus:
> Sulphate of Copper 1,
> Alum 1,
> Nitrate of Potash 1.

Fused together, and Camphor equal to $\frac{1}{50}$ of the whole added.

The preparation is run into moulds to form sticks. It should be kept in a stoppered bottle.

Largely used for the treatment of chronic granular lids.

LEAD.

Lead Lotion :
    Liquor Plumbi Subacetatis (B. P.) ℥j.
    Distilled Water Oj.—(1 in 160.)

Used in chronic conjunctivitis *when the cornea is sound,* and in inflammations of the eyelids and lachrymal sac.

Spirit Lotion :
    Rectified (or Methylated) Spirit ℥iv,
    Water ℥xvj.

Used as an evaporating lotion to allay or prevent inflammation of the wound after operations on the eyelids.

Lead and Spirit Lotion :
    Spirit Lotion Oj,
    Liquor Plumbi Subacetatis (B. P.) ℥ij.

Used in the same cases when there is no fear that the cornea is abraded or ulcerated. A better antiphlogistic than spirit alone.

MERCURY.

Calomel Powder :
Used for dusting on the cornea in some cases of ulceration. Flicked into the eye from a dry camel-hair brush filled with the powder.

Yellow Oxide of Mercury (Yellow Ointment) :
    (1) Unguentum Hydrargyri Oxidi Flavi (B. P.) (1 in 20).
    or Yellow Oxide of Mercury gr. iij.
    Vaseline ℥j (1 in 20).

(2) A weaker preparation, containing gr. j of the Yellow Oxide to ℥j (1 in 60), is sometimes useful.

Used in many cases of corneal ulceration and recent corneal nebulæ, a morsel as large as a hemp-seed being inserted within the lower lid once a day. It is also suitable for ophthalmia tarsi.

Red Oxide of Mercury (Pink Ointment) :
    Red Oxide of Mercury gr. iij.
    Vaseline ℥j.

Used for ophthalmia tarsi, &c. Was formerly used for corneal ulcers and nebulæ; but the Yellow Oxide, which being made by precipitation is not crystalline, is now generally preferred because less irritating.

Nitrate of Mercury (Citrine Ointment) :
Unguentum Hydrargyri Nitratis (B. P.) ʒj,
Vaseline or Prepared Lard ʒvij.
Used in the same cases as the Red Oxide.

SULPHATE OF ZINC :
Sulphate of Zinc gr. j or ij,
Water or Rose Water ʒj.

CHLORIDE OF ZINC :
Chloride of Zinc gr. ij,
Water ʒj,
Dilute Hydrochloric Acid, just enough to
make a clear solution.

ALUM :
Alum gr. iv to gr. x,
Water ʒj.

The above lotions are in common use in the milder forms of
acute and chronic ophthalmia. The Chloride occasionally
irritates; it is especially used in purulent and severe catarrhal
ophthalmia instead of the weak Nitrate of Silver lotions. The
stronger Alum lotion is often used in the same cases. The
Alum and Sulphate of Zinc lotions may be used unsparingly
to the conjunctiva; the Chloride, even in severe cases, not
more than six times a day.

CARBONATE OF SODA :
Carbonate of Soda gr. x,
Water ʒj.

Used for softening the crusts in severe ophthalmia tarsi. A
small quantity of the lotion, diluted with its own bulk of hot
water, to be used for soaking the edges of the eyelids for ten or
fifteen minutes night and morning.

TAR AND SODA :
Carbonate of Soda ʒiss,
Liquor Carbonis Detergens ʒj to ʒss,
Water to Oj.
Used in the same cases as the last.

BORAX :
Biborate of Soda gr. x.
Water ʒj.
Used in the same cases as the last.

ATROPINE:

(1) *Strong Atropine Drops:*

Liquor Atropiæ Sulphatis (B. P.),
(Sulphate of Atropia gr. iv,
Distilled water ℨj.)

Used in all cases where the rapid and full mydriatic action of the drug is required. The ciliary muscle and iris do not thoroughly recover from the effect of complete atropisation for about ten days.

(2) *Weak Atropine Drops:*

Sulphate of Atropia gr. ½,
Distilled water ℨj.

Used when for optical purposes it is desired to keep the pupil partially dilated for a long time, as in immature nuclear cataract. A single drop about three times a week will generally suffice.

Solutions of Sulphate of Atropine will keep for an indefinite time; the flocculent sediment which often forms does not impair their efficiency.

The Liquor Atropiæ (B. P.), which contains rectified spirit, is irritating to the eye and should never be used.

DATURINE:

Sulphate of Daturia gr. iv,
Distilled water ℨj.

Used as a mydriatic in cases where Atropine causes conjunctival irritation.

DUBOISINE:

Sulphate of Duboisia gr. iv,
Distilled water ℨj.

A new mydriatic, acting more quickly and powerfully, and passing off in a shorter time than Atropine. Said to be tolerated in cases where Atropine causes conjunctivitis. To be used with great caution, as well-marked toxic symptoms are often caused even by two or three drops to the eye. A weaker solution (gr. j to ℨj) is safer. It is not yet certain that Duboisine' has any advantages over Atropine.

ESERINE (the Alkaloid of Calabar Bean):

Sulphate of Eseria gr. ij,
Distilled water ℨj.

Used in mydriasis and paralysis of the accommodation whether caused by Atropine or by nerve lesions, in some forms of corneal ulcer, and in acute glaucoma. Its effect only lasts an hour or two at first; after several weeks' use it remains considerably longer, but never nearly so long as that of Atropine.

All the mydriatics and myotics may be obtained in the form of small gelatine discs of known strength, which are sometimes more convenient than the solutions.

BELLADONNA FOMENTATION:

Extract of Belladonna ʒj to ʒij,
Water Oj.

Warmed in a cup or small basin and used as a hot fomentation in suppurating and serpiginous ulcers of cornea.

BANDAGES for the eyes may be of thin flannel or soft calico. A linen or knitted cotton bandage, about ten inches long, with four tails of tape, or a loop of tape embracing the back of the head (Liebreich's bandage), is very convenient after the more serious operations.

When absolute exclusion of light is desired it is best to use a bandage made of a double fold of some thin black material.

Fine old linen is better than lint for placing next the skin in dressing after operations.

SHADES may be made of thin cardboard covered with some dark material, or of stout, dark blue paper, like that used for making grocers' sugar bags. Shades of black plaited straw are also very light and convenient.

Shades, to be effectual, should extend to the temple on each side, so as to exclude all side light.

PROTECTIVE GLASSES:

Various patterns of glasses are made for the purpose of protecting the eyes from wind, dust, and bright light. The glasses are flat, or hollow like a watch glass, and coloured in various shades of blue or smoke tint. The most effectual are the ones known as "goggles;" in these the space between the glass and the edge of the orbit is filled by a carefully fitting framework of fine wire gauze or black muslin, by which side wind and light are excluded.

Other forms, known as "horseshoe" or "D," and "domed" or "hollow," glasses are also in common use.

TEST TYPES:

Snellen's types for testing both near and distant vision under an angle of 5 minutes can be obtained, in several languages, in the form of a small book, from Williams and Norgate, 14, Henrietta Street, Covent Garden.

The types which I generally use for testing near vision are those used at the Moorfields Hospital, where they may be obtained. They can also be bought, conveniently mounted, of

Walters and Co., Instrument Makers, Albert Embankment, S.E. These types nearly resemble those of Jaeger, and, though less correct theoretically than the corresponding types of Snellen's scale, are more convenient in practice for testing the reading power.

A convenient set of tests, small enough to be carried in the pocket, has been arranged for me by Mr. Hawksley, 300, Oxford Street. It consists of types for near and distant vision, a pupilometer for measuring the pupil, a set of coloured stuffs for colour blindness, and a small series of lenses for testing refraction. This case is intended chiefly for ward work and general medical cases. It may be also bought without the lenses.

The set of COLOURED WOOLS recommended by Prof. Holm-gren, of Upsala, for testing colour blindness can be obtained for about five and sixpence from P. Dörffel, Unter den Linden 46, Berlin.

INDEX

ERRATA.

Pp. 25 to 42, head-line : *for* "ophthalmic examination," *read* "ophthalmoscopic examination."

P. 41, *for* " outer layer shaded," *read* " outer layers shaded," in description of Fig. 9, line 2.

P. 232, line 27, *for* (p. 224) *read* (p. 225).

ADDENDA.

Perception of Light.—P. 4, *add* to Abbreviations : " p. l. = perception of light; perception of the difference between strong light and shade when the eye is alternately covered and exposed."

Optic Axis, and Visual Axis.—P. 250, line 4, *for* "joining the y. s. with the centre of the cornea," *read* "joining the y. s. with a point close to, but usually a little to the inner side of, the centre . . ." The *optic* axis does not quite coincide with the *visual* axis or visual line, but passes from the centre of the cornea through the optical centre of the eye to a point lying between the y. s. and o. d.

The Teeth in Hereditary Syphilis.—None of the *first set* of teeth are characteristically altered, though the incisors frequently decay early.

In the *permanent set* only two teeth, the central upper incisors, are to be relied upon; but the other incisors, both upper and lower, and the first molars, are often deformed from the same cause. The characteristic change in the upper central incisors appears to depend upon defective formation of the dentine, and in a less degree of the enamel, of the central lobe of the tooth. Soon after the eruption of the tooth this lobe wears away, leaving at the centre of the cutting edge a vertical notch. If the change have been so great as to prevent the formation of the central lobe, we find instead of the notch, a narrowing and thinning of the cutting edge in comparison with the crown, which, according to its degree, produces a resemblance to a screw-driver, or to a peg. The teeth are also usually too small in every dimension, so that the incisors are often separated from one another by considerable spaces. In extreme cases all the incisors are peggy and much dwarfed.

PRINTED BY
J. E. ADLARD, BARTHOLOMEW CLOSE

J. & A. CHURCHILL'S

MEDICAL CLASS BOOKS.

ANATOMY.

BRAUNE.—An Atlas of Topographical Anatomy, after Plane Sections of Frozen Bodies. By WILHELM BRAUNE, Professor of Anatomy in the University of Leipzig. Translated by EDWARD BELLAMY, F.R.C.S., Surgeon to Charing-Cross Hospital, and Lecturer on Anatomy in its School. With 34 Photo-lithographic Plates and 46 Woodcuts. Large imp. 8vo, 40s.

FLOWER.—Diagrams of the Nerves of the Human Body, exhibiting their Origin, Divisions, and Connexions, with their Distribution to the various Regions of the Cutaneous Surface, and to all the Muscles. By WILLIAM H. FLOWER, F.R.C.S., F.R.S., Conservator of the Museum of the Royal College of Surgeons. Second Edition, containing 6 Plates. Royal 4to, 12s.

GODLEE.—An Atlas of Human Anatomy: illustrating most of the ordinary Dissections; and many not usually practised by the Student. By RICKMAN J. GODLEE, M.S., F.R.C.S., Assistant-Surgeon to University College Hospital, and Senior Demonstrator of Anatomy in University College. To be completed in 12 or 13 Bi-monthly Parts, each containing 4 Coloured Plates, with Explanatory Text. Parts I to X. Imp. 4to, 7s. 6d. each.

HEATH.—Practical Anatomy: a Manual of Dissections. By CHRISTOPHER HEATH, F.R.C.S., Holme Professor of Clinical Surgery in University College and Surgeon to the Hospital. Fourth Edition. With 16 Coloured Plates and 264 Engravings. Crown 8vo, 14s.

NEW BURLINGTON STREET.

J. & A. Churchill's Medical Class Books.

ANATOMY—continued.

HOLDEN.—Human Osteology: comprising a Description of the Bones. with Delineations of the Attachments of the Muscles, the General and Microscopical Structure of Bone and its Development. By LUTHER HOLDEN, F.R.C.S., Senior Surgeon to St. Bartholomew's and the Foundling Hospitals, and ALBAN DORAN, F.R.C.S., late Anatomical, now Pathological, Assistant to the Museum of the Royal College of Surgeons. Fifth Edition. With 61 Lithographic Plates and 89 Engravings. Royal 8vo, 16s.

By the same Author.

A Manual of the Dissection of the Human Body. Fourth Edition. Revised by the Author and JOHN LANGTON, F.R.C.S., Assistant Surgeon and Lecturer on Anatomy at St. Bartholomew's Hospital. With Engravings. 8vo, 16s.

ALSO,

Landmarks, Medical and Surgical. Second Edition. 8vo, 3s. 6d.

MORRIS.—The Anatomy of the Joints of Man. By HENRY MORRIS, M.A., F.R.C.S., Surgeon to, and Lecturer on Anatomy and Practical Surgery at, the Middlesex Hospital. With 44 Plates (19 Coloured) and Engravings. 8vo, 16s.

WAGSTAFFE.—The Student's Guide to Human Osteology. By WM. WARWICK WAGSTAFFE, F.R.C.S., Assistant Surgeon to, and Lecturer on Anatomy at, St. Thomas's Hospital With 23 Plates and 66 Engravings. Fcap. 8vo, 10s. 6d.

WILSON.—The Anatomist's Vade-Mecum: a System of Human Anatomy. By ERASMUS WILSON, F.R.C.S., F.R.S. late Professor of Dermatology to the Royal College of Surgeons Ninth Edition, by G. BUCHANAN, M.A., M.D., late Professor of Clinical Surgery in the University of Glasgow, and HENRY E. CLARK, F.F.P.S. Lecturer on Anatomy in the Glasgow Royal Infirmary School o Medicine. With 371 Engravings. Crown 8vo, 14s.

Anatomical Remembrancer (the); or, Complete Pocket Anatomist. Eighth Edition. 32mo, 3s. 6d.

BOTANY.

BENTLEY.—A Manual of Botany. By Robert

BENTLEY, F.L.S., Professor of Botany in King's College and to the Pharmaceutical ' Society. With 1138 Engravings. Third Edition. Crown 8vo, 14s.

BENTLEY AND TRIMEN.—Medicinal Plants:

being descriptions, with original Figures, of the Principal Plants employed in Medicine, and an account of their Properties and Uses By ROBERT BENTLEY, F.L.S., and HENRY TRIMEN, M.B., F.L.S., British Museum, and Lecturer on Botany at St. Mary's Hospital Medical School. To be completed in 42 Monthly Parts, each containing 8 Coloured Plates. Parts I. to XL. Large 8vo, 5s. each part.

CHEMISTRY.

BERNAYS.—Notes for Students in Chemistry;

being a Syllabus of Chemistry compiled mainly from the Manuals of Fownes-Watts, Miller, Wuiz, and Schorlemmer. By ALBERT J. BERNAYS, Ph.D., Professor of Chemistry at St. Thomas's Hospital, Examiner in Chemistry at the Royal College of Physicians of London. Sixth Edition. Fcap. 8vo, 3s. 6d.

*By the same Author.

Skeleton Notes on Analytical Chemistry,

for Students in Medicine. Fcap. 8vo, 2s. 6d.

BLOXAM.—Chemistry, Inorganic and Organic;

with Experiments. By CHARLES L. BLOXAM, Professor of Chemistry in King's College. Third Edition. With 295 Engravings. 8vo, 16s.

By the same Author.

Laboratory Teaching; or, Progressive

Exercises in Practical Chemistry. Fourth Edition. With 83 Engravings. Crown 8vo, 5s. 6d.

BOWMAN AND BLOXAM.—Practical Chemistry,

including Analysis. By JOHN E. BOWMAN, Formerly Professor of Practical Chemistry in King's College, and CHARLES L. BLOXAM, Professor of Chemistry in King's College. With 98 Engravings. Seventh Edition. Fcap. 8vo, 6s. 6d.

CHEMISTRY—*continued*.

CLOWES.—Practical Chemistry and Qualitative Inorganic Analysis. An Elementary Treatise specially adapted for use in the Laboratories of Schools and Colleges, and by Beginners. By FRANK CLOWES, D.SC., Senior Science Master at the High School, Newcastle-under-Lyme. Second Edition. With 47 Engravings. Post 8vo, 7s. 6d.

FOWNES AND WATTS.—Physical and Inorganic Chemistry. Twelfth Edition. By GEORGE FOWNES, F.R.S., and HENRY WATTS, B.A., F.R.S. With 154 Engravings, and Coloured Plate of Spectra. Crown 8vo, 8s. 6d.

By the same Authors.

Chemistry of Carbon - Compounds, or Organic Chemistry. Twelfth Edition. With Engravings. Crown 8vo, 10s.

GALLOWAY.—A Manual of Qualitative Analysis. By ROBERT GALLOWAY, Professor of Applied Chemistry in the Royal College of Science for Ireland. Fifth Edition. With Engravings. Post 8vo, 8s. 6d.

VACHER.—A Primer of Chemistry, including Analysis. By ARTHUR VACHER. 18mo, 1s.

VALENTIN.—Introduction to Inorganic Chemistry. By WILLIAM G. VALENTIN, F.C.S., Late Principal Demonstrator of Practical Chemistry in the Science Training Schools. Third Edition. With 82 Engravings. 8vo, 6s. 6d.

By the same Author.

A Course of Qualitative Chemical Analysis. Fourth Edition. With 19 Engravings. 8vo, 7s. 6d.

ALSO,

Chemical Tables for the Lecture-room and Laboratory. In Five large Sheets, 5s. 6d.

CHILDREN, DISEASES OF.

ELLIS.—A Practical Manual of the Diseases
of Children. By EDWARD ELLIS, M.D., late Senior Physician to the
Victoria Hospital for Sick Children. With a Formulary. Third
Edition. Crown 8vo, 7s. 6d.

SMITH. — Clinical Studies of Disease in
Children. By EUSTACE SMITH, M.D., F.R.C.P., Physician to H.M. the
King of the Belgians, and to the East London Hospital for Children.
Post 8vo, 7s. 6d.

By the same Author.

On the Wasting Diseases of Infants and
Children. Third Edition. Post 8vo, 8s. 6d.

STEINER.—Compendium of Children's Dis-
eases; a Handbook for Practitioners and Students. By JOHANN
STEINER, M.D. Translated from the Second German Edition, by LAWSON
TAIT, F.R.C.S., Surgeon to the Birmingham Hospital for Women, &c.
8vo, 12s. 6d.

DENTISTRY.

SEWILL.—The Student's Guide to Dental
Anatomy and Surgery. By HENRY E. SEWILL, M.R.C.S., L.D.S., late
Dental Surgeon to the West London Hospital. With 77 Engravings.
Fcap. 8vo, 5s. 6d.

SMITH.—Handbook of Dental Anatomy and
Surgery. For the Use of Students and Practitioners. By JOHN SMITH,
M.D., F.R.S.E., Dental Surgeon to the Royal Infirmary, Edinburgh.
Second Edition. Fcap. 8vo, 4s. 6d.

STOCKEN.—Elements of Dental Materia Medica
and Therapeutics, with Pharmacopœia. By JAMES STOCKEN, L.D.S.R.C.S.,
Lecturer on Dental Materia Medica and Therapeutics and Dental
Surgeon to the National Dental Hospital. Second Edition. Fcap. 8vo,
6s. 6d.

NEW BURLINGTON STREET.

DENTISTRY—*continued.*

TAFT.—A Practical Treatise on Operative Dentistry. By JONATHAN TAFT, D.D.S., Professor of Operative Surgery in the Ohio College of Dental Surgery. Third Edition. With 134 Engravings. 8vo, 18s.

TOMES (C. S.).—Manual of Dental Anatomy, Human and Comparative. By CHARLES S. TOMES, M.A., M.R.C.S., Lecturer on Anatomy and Physiology at the Dental Hospital of London. With 179 Engravings. Crown 8vo, 10s. 6d.

TOMES (J. and C. S.).—A Manual of Dental Surgery. By JOHN TOMES, M.R.C.S., F.R.S., Consulting Surgeon-Dentist to Middlesex Hospital; and CHARLES S. TOMES, M.A., M.R.C.S., Lecturer on Anatomy and Physiology at the Dental Hospital of London. Second Edition. With 262 Engravings. Fcap. 8vo, 14s.

EAR, DISEASES OF.

BURNETT.—The Ear: its Anatomy, Physiology, and Diseases. A Practical Treatise for the Use of Medical Students and Practitioners. By CHARLES H. BURNETT, M.D., Aural Surgeon to the Presbyterian Hospital, Philadelphia. With 87 Engravings. 8vo, 18s.

DALBY.—On Diseases and Injuries of the Ear. By WILLIAM B. DALBY, F.R.C.S., Aural Surgeon to, and Lecturer on Aural Surgery at, St. George's Hospital. With Engravings. Fcap. 8vo, 6s. 6d.

JONES.—A Practical Treatise on Aural Surgery. By H. MACNAUGHTON JONES, M.D., Professor of the Queen's University in Ireland, Surgeon to the Cork Ophthalmic and Aural Hospital. With 46 Engravings. Crown 8vo, 5s.

By the same Author.

Atlas of the Diseases of the Membrana Tympani. In Coloured Plates, containing 59 Figures. With Explanatory Text. Crown 4to, 21s.

J. & A. Churchill's Medical Class Books.

FORENSIC MEDICINE.

OGSTON.—Lectures on Medical Jurisprudence.
By FRANCIS OGSTON, M.D., Professor of Medical Jurisprudence and Medical Logic in the University of Aberdeen. Edited by FRANCIS OGSTON, Jun., M.D., Assistant to the Professor of Medical Jurisprudence and Lecturer on Practical Toxicology in the University of Aberdeen. With 12 Plates. 8vo, 18s.

TAYLOR.—The Principles and Practice of
Medical Jurisprudence. By ALFRED S. TAYLOR, M.D., F.R.S., Professor of Medical Jurisprudence to Guy's Hospital. Second Edition. With 189 Engravings. 2 Vols. 8vo, 31s. 6d.

By the same Author.

A Manual of Medical Jurisprudence.
Tenth Edition. With 55 Engravings. Crown 8vo, 14s.

ALSO,

On Poisons, in relation to Medical Juris-
prudence and Medicine. Third Edition. With 104 Engravings. Crown 8vo, 16s.

WOODMAN AND TIDY.—A Handy-Book of
Forensic Medicine and Toxicology. By W. BATHURST WOODMAN, M.D., F.R.C.P. ; and C. MEYMOTT TIDY, M.B., Professor of Chemistry and of Medical Jurisprudence, &c., at the London Hospital. With 8 Lithographic Plates and 116 Wood Engravings. 8vo, 31s. 6d.

HYGIENE.

WILSON.—A Handbook of Hygiene and Sani-
tary Science. By GEORGE WILSON, M.A., M.D., Medical Officer of Health for Mid Warwickshire. Fourth Edition. With Engravings. Post 8vo, 10s. 6d.

NEW BURLINGTON STREET.

7

J. & A. Churchill's Medical Class Books.

PARKES.—A Manual of Practical Hygiene.

By EDMUND A. PARKES, M.D., F.R.S. Fifth Edition by F. DE CHAUMONT, M.D., F.R.S., Professor of Military Hygiene in the Army Medical School. With 9 Plates and 112 Engravings. 8vo, 18s.

By the same Author.

Public Health: being a Concise Sketch of

the Sanitary Considerations connected with the Land, with Cities, Villages, Houses, and Individuals. Revised by WILLIAM AITKEN, M.D., F.R.S., Professor of Pathology in the Army Medical School. Crown 8vo, 2s. 6d.

MATERIA MEDICA AND THERAPEUTICS.

BINZ AND SPARKS.—The Elements of Thera-

peutics: a Clinical Guide to the Action of Medicines. By C. BINZ, M.D., Professor of Pharmacology in the University of Bonn. Translated from the Fifth German Edition, and Edited with Additions, in conformity with the British and American Pharmacopœias, by EDWARD I. SPARKS, M.A., M.B. Oxon., F.R.C.P. Lond. Crown 8vo, 8s. 6d.

ROYLE AND HARLEY.—A Manual of Materia

Medica and Therapeutics. By J. FORBES ROYLE, M.D., F.R.S., formerly Professor of Materia Medica in King's College; and JOHN HARLEY, M.D., F.R.C.P., Physician to, and Joint Lecturer on Clinical Medicine at, St. Thomas's Hospital. Sixth Edition. With 139 Engravings. Crown 8vo, 15s.

THOROWGOOD. — The Student's Guide to

Materia Medica. By JOHN C. THOROWGOOD, M.D., F.R.C.P., Lecturer on Materia Medica at the Middlesex Hospital. With Engravings. Fcap. 8vo, 6s. 6d.

WARING.—A Manual of Practical Therapeu-

tics. By EDWARD J. WARING, M.D., F.R.C.P., Retired Surgeon H.M. Indian Army. Third Edition. Fcap. 8vo, 12s. 6d.

MEDICINE.

BARCLAY.—A Manual of Medical Diagnosis.
By A. WHYTE BARCLAY, M.D., F.R.C.P., Physician to, and Lecturer on Medicine at, St. George's Hospital. Third Edition. Fcap. 8vo, 10s. 6d.

BARLOW.—A Manual of the Practice of Medicine. By HILARO BARLOW, M.D., Formerly Senior Physician to Guy's Hospital. Second Edition. Fcap. 8vo, 7s. 6d.

CHARTERIS.—The Student's Guide to the Practice of Medicine. By MATTHEW CHARTERIS, M.D., Professor of Practice of Medicine, Anderson's College; Physician and Lecturer on Clinical Medicine, Royal Infirmary, Glasgow. With Engravings on Copper and Wood. Second Edition. Fcap. 8vo, 6s. 6d.

FENWICK.—The Student's Guide to Medical Diagnosis. By SAMUEL FENWICK, M.D., F.R.C.P., Physician to the London Hospital. Fourth Edition. With 106 Engravings. Fcap. 8vo, 6s. 6d.

By the same Author.

The Student's Outlines of Medical Treatment. Fcap. 8vo., 7s.

FLINT.—Clinical Medicine : a Systematic Treatise on the Diagnosis and Treatment of Disease. By AUSTIN FLINT, M.D., Professor of the Principles and Practice of Medicine, &c., in Bellevue Hospital Medical College. 8vo, 20s.

By the same Author. •

A Manual of Percussion and Auscultation ; of the Physical Diagnosis of Diseases of the Lungs and Heart, and of Thoracic Aneurism. Post 8vo, 6s. 6d.

HALL.—Synopsis of the Diseases of the Larynx, Lungs, and Heart : comprising Dr. Edwards' Tables on the Examination of the Chest. With Alterations and Additions. By F. DE HAVILLAND HALL, M.D., Assistant-Physician to the Westminster Hospital. Royal 8vo, 2s. 6d.

MIDWIFERY.

BARNES.—Lectures on Obstetric Operations, including the Treatment of Hæmorrhage, and forming a Guide to the Management of Difficult Labour. By ROBERT BARNES, M.D., F.R.C.P., Obstetric Physician to, and Lecturer on Diseases of Women, &c., at St. George's Hospital. Third Edition. With 124 Engravings. 8vo, 18s.

CLAY.—The Complete Handbook of Obstetric Surgery; or, Short Rules of Practice in every Emergency, from the Simplest to the most formidable Operations connected with the Science of Obstetricy. By CHARLES CLAY, M.D., late Senior Surgeon to, and Lecturer on Midwifery at, St. Mary's Hospital, Manchester. Third Edition. With 91 Engravings. Fcap. 8vo, 6s. 6d.

RAMSBOTHAM.—The Principles and Practice of Obstetric Medicine and Surgery. By FRANCIS H. RAMSBOTHAM, M.D., formerly Obstetric Physician to the London Hospital. Fifth Edition. Illustrated with 120 Plates, forming one thick handsome volume. 8vo, 22s.

ROBERTS.—The Student's Guide to the Practice of Midwifery. By D. LLOYD ROBERTS, M.D., F.R.C.P., Physician to St. Mary's Hospital, Manchester. Second Edition. With 96 Engravings. Fcap. 8vo, 7s.

SCHROEDER.—A Manual of Midwifery; including the Pathology of Pregnancy and the Puerperal State. By KARL SCHROEDER, M.D., Professor of Midwifery in the University of Erlangen. Translated by CHARLES H. CARTER, M.D. With Engravings. 8vo, 12s. 6d.

SWAYNE.—Obstetric Aphorisms for the Use of Students commencing Midwifery Practice. By JOSEPH G. SWAYNE, M.D., Lecturer on Midwifery at the Bristol School of Medicine. Sixth Edition. With Engravings. Fcap. 8vo, 3s. 6d.

NEW BURLINGTON STREET.

MICROSCOPY.

CARPENTER.—The Microscope and its Revela-
tions. By WILLIAM B. CARPENTER, C.B., M.D., F.R.S., late Registrar
to the University of London. Fifth Edition. With more than 500
Engravings. Crown 8vo, 15s.

MARSH.—Section-Cutting : a Practical Guide
to the Preparation and Mounting of Sections for the Microscope, special
prominence being given to the subject of Animal Sections. By Dr.
SYLVESTER MARSH. With Engravings. Fcap. 8vo, 2s. 6d.

MARTIN.—A Manual of Microscopic Mounting.
By JOHN H. MARTIN, Member of the Society of Public Analysts, &c.
Second Edition. With several Plates and 144 Engravings. 8vo, 7s. 6d.

WYTHE.—The Microscopist : a Manual of
Microscopy and Compendium of the Microscopic Sciences, Micro-
Mineralogy, Micro-Chemistry, Biology, Histology, and Pathological
Histology. By J. H. WYTHE, A.M., M.D., Professor of Microscopy and
Biology in the San Francisco Medical College. Third Edition. With
205 Illustrations. Royal 8vo, 18s.

OPHTHALMOLOGY.

HIGGENS.—Hints on Ophthalmic Out-Patient
Practice. By CHARLES HIGGENS, F.R.C.S., Ophthalmic Assistant-Sur-
geon to, and Lecturer on Ophthalmology at, Guy's Hospital. Second
Edition. Fcap. 8vo, 3s.

JONES.—A Manual of the Principles and
Practice of Ophthalmic Medicine and Surgery. By T. WHARTON JONES,
F.R.C.S., F.R.S., Ophthalmic Surgeon and Professor of Ophthalmology
to University College Hospital. Third Edition. With 9 Coloured
Plates and 173 Engravings. Fcap. 8vo, 12s. 6d.

MACNAMARA.—A Manual of the Diseases of
the Eye. By CHARLES MACNAMARA, F.R.C.S., Surgeon to Westminster
Hospital. Third Edition. With 7 Coloured Plates and 52 Engravings.
Fcap. 8vo, 12s. 6d.

OPHTHALMOLOGY—*continued.*

NETTLESHIP.—The Student's Guide to Diseases
of the Eye. By EDWARD NETTLESHIP, F.R.C.S., Ophthalmic Surgeon
to, and Lecturer on Ophthalmic Surgery at, St. Thomas's Hospital.
With 48 Engravings. Fcap. 8vo, 7s. 6d.

WELLS.—A Treatise on the Diseases of the
Eye. By J. SOELBERG WELLS, F.R.C.S., Ophthalmic Surgeon to King's
College Hospital, Professor of Ophthalmology at King's College. With
Coloured Plates and Engravings. Third Edition. 8vo, 25s.

PATHOLOGY.

JONES AND SIEVEKING.—A Manual of Patho-
logical Anatomy. By C. HANDFIELD JONES, M.B., F.R.S., Physician to
St. Mary's Hospital, and EDWARD H. SIEVEKING, M.D., F.R.C.P., Physi-
cian to St. Mary's Hospital. Second Edition. Edited by J. F. PAYNE,
M.B., Assistant-Physician and Lecturer on General Pathology at St.
Thomas's Hospital. With 195 Engravings. Crown 8vo, 16s.

VIRCHOW. — Post-Mortem Examinations: a
Description and Explanation of the Method of Performing them,
with especial reference to Medico-Legal Practice. By Professor
RUDOLPH VIRCHOW, Berlin Charité Hospital. Fcap. 8vo, 2s. 6d.

WILKS AND MOXON.—Lectures on Pathologi-
cal Anatomy. By SAMUEL WILKS, M.D., F.R.S., Physician to, and
Lecturer on Medicine at, Guy's Hospital; and WALTER MOXON, M.D.,
F.R.C.P., Physician to, and Lecturer on Clinical Medicine at, Guy's
Hospital. Second Edition. With 5 Steel Plates. 8vo, 18s.

PSYCHOLOGY.

BUCKNILL AND TUKE.—A Manual of Psycho-
logical Medicine: containing the Lunacy Laws, Nosology, Ætiology,
Statistics, Description, Diagnosis, Pathology, and Treatment of Insanity,
with an Appendix of Cases. By JOHN C. BUCKNILL, M.D., F.R.S.,
and D. HACK TUKE, M.D., F.R.C.P. Fourth Edition, with 12 Plates
(30 Figures). 8vo, 25s.

J. & A. Churchill's Medical Class Books.

PHYSIOLOGY.

CARPENTER.—**Principles of Human Physio-**
logy. With 3 Steel Plates and 371 Engravings. By WILLIAM B.
CARPENTER, C.B., M.D., F.R.S., late Registrar to the University of
London. Eighth Edition. Edited by Mr. Henry Power. 8vo, 31s. 6d.

By the same Author.

A Manual of Physiology. With upwards
of 250 Illustrations. Fifth Edition. Edited by P. H. PYE-SMITH,
M.D., F.R.C.P. Crown 8vo. [*In the press.*

DALTON.—**A Treatise on Human Physiology :**
designed for the use of Students and Practitioners of Medicine. By
JOHN C. DALTON, M.D., Professor of Physiology and Hygiene in the
College of Physicians and Surgeons, New York. Sixth Edition. With
316 Engravings. Royal 8vo, 20s.

FREY.—**The Histology and Histo-Chemistry of**
Man. A Treatise on the Elements of Composition and Structure of the
Human Body. By HEINRICH FREY, Professor of Medicine in Zurich.
Translated from the Fourth German Edition, by ARTHUR E. BARKER,
Assistant-Surgeon to the University College Hospital. With 608
Engravings. 8vo, 21s.

RUTHERFORD.—**Outlines of Practical Histo-**
logy. By WILLIAM RUTHERFORD, M.D., F.R.S., Professor of the Insti-
tutes of Medicine in the University of Edinburgh ; Examiner in
Physiology in the University of London. Second Edition. With 63
Engravings. Crown 8vo (with additional leaves for Notes), 6s.

SANDERSON.—**Handbook for the Physiological**
Laboratory: containing an Exposition of the fundamental facts of the
Science, with explicit Directions for their demonstration. By J. BURDON
SANDERSON, M.D., F.R.S., Professor and Superintendent of the Brown
Institution; E. KLEIN, M.D., F.R.S., Assistant-Professor in the Brown
Institution ; MICHAEL FOSTER, M.D., F.R.S., Prælector of Physiology
at Trinity College, Cambridge ; and T. LAUDER BRUNTON, M.D., F.R.S.,
Lecturer on Materia Medica at St. Bartholomew's Hospital Medical
College. 2 Vols., with 123 Plates. 8vo, 24s.

SURGERY.

BRYANT. — A Manual for the Practice of
Surgery. By THOMAS BRYANT, F.R.C.S., Surgeon to, and Lecturer on
Surgery at, Guy's Hospital. Third Edition: With 672 Engravings
(nearly all original, many being coloured). 2 vols. Crown 8vo, 28s.

BELLAMY. — The Student's Guide to Surgical
Anatomy; a Text-Book for the Pass Examination. By EDWARD
BELLAMY, F.R.C.S., Surgeon to, and Lecturer on Anatomy and Practical
Surgery at, Charing Cross Hospital. With 50 Engravings. Fcap. 8vo,
6s. 6d.

CLARK AND WAGSTAFFE. — Outlines of
Surgery and Surgical Pathology. By F. LE GROS CLARK, F.R.C.S.,
F.R.S., Consulting Surgeon to St. Thomas's and the Great Northern
Hospitals. Second Edition. Revised and expanded by the Author,
assisted by W. W. WAGSTAFFE, F.R.C.S., Assistant-Surgeon to St.
Thomas's Hospital. 8vo, 10s. 6d.

DRUITT. — The Surgeon's Vade-Mecum; a
Manual of Modern Surgery. By ROBERT DRUITT, F.R.C.S. Eleventh
Edition. With 369 Engravings. Fcap. 8vo, 14s.

FERGUSSON. — A System of Practical Surgery.
By Sir WILLIAM FERGUSSON, Bart., F.R.C.S., F.R.S., late Surgeon and
Professor of Clinical Surgery to King's College Hospital. With 463
Engravings. Fifth Edition. 8vo, 21s.

HEATH. — A Manual of Minor Surgery and
Bandaging, for the use of House-Surgeons, Dressers, and Junior Practi-
tioners. By CHRISTOPHER HEATH, F.R.C.S., Holme Professor of Clinical
Surgery in University College and Surgeon to the Hospital. Fifth
Edition. With 83 Engravings. Fcap. 8vo, 5s. 6d.

By the same Author.

A Course of Operative Surgery: with
Twenty Plates drawn from Nature by M. LÉVEILLÉ, and Coloured
by hand under his direction. Large 8vo, 40s.

ALSO,

The Student's Guide to Surgical Diagnosis. Fcap. 8vo, 6s. 6d.

SURGERY—*continued.*

MAUNDER.—Operative Surgery. By Charles
F. MAUNDER, F.R.C.S., late Surgeon to, and Lecturer on Surgery at,
the London Hospital. Second Edition. With 164 Engravings. Post
8vo, 6s.

PIRRIE.—The Principles and Practice of
Surgery. By WILLIAM PIRRIE, F.R.S.E., Professor of Surgery in the
University of Aberdeen. Third Edition. With 490 Engravings. 8vo, 28s.

TERMINOLOGY.

DUNGLISON.—Medical Lexicon: a Dictionary
of Medical Science, containing a concise Explanation of its various
Subjects and Terms, with Accentuation, Etymology, Synonymes, &c.
By ROBLEY DUNGLISON, M.D. New Edition, thoroughly revised by
RICHARD J. DUNGLISON, M.D. Royal 8vo, 28s.

MAYNE.—A Medical Vocabulary: being an
Explanation of all Terms and Phrases used in the various Depart-
ments of Medical Science and Practice, giving their Derivation, Meaning,
Application, and Pronunciation. By ROBERT G. MAYNE, M.D., LL.D.,
and JOHN MAYNE, M.D., L.R.C.S.E. Fourth Edition. Fcap. 8vo, 10s.

WOMEN, DISEASES OF.

BARNES.—A Clinical History of the Medical
and Surgical Diseases of Women. By ROBERT BARNES, M.D., F.R.C.P.,
Obstetric Physician to, and Lecturer on Diseases of Women, &c., at, St.
George's Hospital. Second Edition. With 181 Engravings. 8vo, 28s.

DUNCAN.—Clinical Lectures on Diseases of
Women. By J. MATTHEWS DUNCAN, M.D., Obstetric Physician to St.
Bartholomew's Hospital. 8vo.

EMMET. — The Principles and Practice of
Gynæcology. By THOMAS ADDIS EMMET, M.D., Surgeon to the
Woman's Hospital of the State of New York. With 130 Engravings.
Royal 8vo, 24s.

WOMEN, DISEASES OF—*continued.*

GALABIN.—The Student's Guide to the Diseases of Women. By ALFRED L. GALABIN, M.D., F.R.C.P., Assistant Obstetric Physician and Joint Lecturer on Obstetric Medicine to Guy's Hospital. With 63 Engravings. Fcap. 8vo, 7s. 6d.

SMITH.—Practical Gynæcology: a Handbook of the Diseases of Women. By HEYWOOD SMITH, M.D., Physician to the Hospital for Women, and to the British Lying-in Hospital. With Engravings. Crown 8vo, 5s. 6d.

WEST AND DUNCAN.—Lectures on the Diseases of Women. By CHARLES WEST, M.D., F.R.C.P. Fourth Edition. Revised and in part re-written by the Author, with numerous additions, by J. MATTHEWS DUNCAN, M.D., Obstetric Physician to St. Bartholomew's Hospital. 8vo, 16s.

ZOOLOGY.

BRADLEY.—Manual of Comparative Anatomy and Physiology. By S. MESSENGER BRADLEY, F.R.C.S., Lecturer on Practical Surgery in Owen's College, Manchester. Third Edition. With 61 Engravings. Post 8vo, 6s. 6d.

CHAUVEAU AND FLEMING.—The Comparative Anatomy of the Domesticated Animals. By A. CHAUVEAU, Professor at the Lyons Veterinary School; and GEORGE FLEMING, Veterinary Surgeon, Royal Engineers. With 450 Engravings. 8vo, 31s. 6d.

HUXLEY.—Manual of the Anatomy of Invertebrated Animals. By THOMAS H. HUXLEY, LL.D., F.R.S. With 156 Engravings. Fcap. 8vo, 16s.

By the same Author.

Manual of the Anatomy of Vertebrated Animals. With 110 Engravings. Post 8vo, 12s.

WILSON.—The Student's Guide to Zoology: a Manual of the Principles of Zoological Science. By ANDREW WILSON, Lecturer on Natural History, Edinburgh. With Engravings. Fcap. 8vo, 6s. 6d.